Elementary Cosmology:
From Aristotle's Universe to the Big Bang and Beyond

Elementary Cosmology: From Aristotle's Universe to the Big Bang and Beyond

James J Kolata

University of Notre Dame, USA

Morgan & Claypool Publishers

Rights & Permissions
To obtain permission to re-use copyrighted material from Morgan & Claypool Publishers, please contact info@morganclaypool.com.

ISBN   978-1-6817-4100-0 (ebook)
ISBN   978-1-6817-4036-2 (print)
ISBN   978-1-6817-4228-1 (mobi)

DOI   10.1088/978-1-6817-4100-0

Version: 20151101

IOP Concise Physics
ISSN 2053-2571 (online)
ISSN 2054-7307 (print)

A Morgan & Claypool publication as part of IOP Concise Physics
Published by Morgan & Claypool Publishers, 40 Oak Drive, San Rafael, CA, 94903, USA

IOP Publishing, Temple Circus, Temple Way, Bristol BS1 6HG, UK

To my wife Ann who encouraged me throughout this book project, and to the many Notre Dame students whose questions and comments over the years contributed immensely to the development of the Elementary Cosmology course.

Contents

Preface

Cosmology is the study of the origin, size, and evolution of the entire Universe. Every culture has developed a cosmology, whether it be based on religious, philosophical, or scientific principles. In this book, the evolution of the scientific understanding of the Universe in the Western tradition is traced from the early Greek philosophers to the most modern 21st century view.

This book began as a series of lecture notes for a one-semester course at the University of Notre Dame called 'Elementary Cosmology'. An elective for non-science majors, it was designed to acquaint the non-mathematically inclined student with the most important discoveries in cosmology up to the present day and how they have constantly altered our perceptions of the origin and structure of the Universe. It examined such questions as 'Where did the Universe come from?', 'Why do scientists now feel sure that its birth was in a great cosmic fireball called the Big Bang?', and 'Where did the Big Bang itself come from?'. The emphasis was on class discussion of readings from 'science popularizations' for the curious and intelligent layperson, focusing eventually on the many interesting and exciting new discoveries in cosmology in the late 20th and early 21st century.

After a brief introduction to the concept of the scientific method, which underpins all scientific approaches to the study of the Universe, the first part of the book describes the way in which detailed observations of the Universe, first with the naked eye and later with increasingly complex modern instruments, ultimately led to the development of the Big Bang theory which supplies the framework for our current understanding of cosmology. The key to this theory is the realization that our Universe, far from being static and eternal, has instead been expanding in size since its origin some 13.5 billion years ago. While the fact of this expansion was accepted rather soon after it was first proposed in the 1920s and 1930s, the more radical idea that the Universe had a birthday was more difficult for scientists to accept. It was only with the development of modern, satellite-based communication devices in the 1960s that instruments sensitive enough to detect the cosmic microwave background (CMB) were produced. It is now understood that the CMB consists of ancient light, emitted at a time near to the formation of the Universe. As such it is the 'smoking gun' of the Big Bang and detailed studies of its properties have led to many interesting and fascinating new discoveries in cosmology.

The second part of the book traces the evolution of the Big Bang theory itself, including the very recent observation that the expansion of the Universe is itself accelerating with time. In addition, the contribution of modern physics to our understanding of the mechanism of the Big Bang is discussed and the state of the Universe at various eras throughout its history is described. Finally, some speculations beyond current knowledge that bear on cosmology are introduced and the implications for future developments in our understanding of the Universe are described.

The text contains many links to websites that clarify and extend the discussion. By following these links (some to images and videos), the reader can attain a much more in-depth understanding of many of the concepts introduced in this book. This is especially true for those seeking a more mathematical discussion of the topics, which is beyond the level of the current text.

Acknowledgements

I acknowledge with gratitude the assistance of Karen Donnison, and Jacky Mucklow at IOP Publishing in production of the manuscript into its final form.

Author biography

James J Kolata

 James J Kolata is an emeritus professor of physics at the University of Notre Dame in the USA and a Fellow of the American Physical Society. He is the author of over 250 research publications in nuclear physics.

Elementary Cosmology: From Aristotle's Universe
to the Big Bang and Beyond

James J Kolata

Chapter 1

The scientific method

1.1 Introduction to the scientific method

The *scientific method* is a procedure for obtaining knowledge about the Universe around us. It begins with the assumption that there is an objective reality that can be observed and measured, and provides a framework for systematizing and extending these observations. A key characteristic of the method is that it seeks to eliminate individual bias by insisting on reproducibility of results. In order to describe the method in more detail, it is necessary to define a number of terms, some of which have different meanings from the way they are understood in common usage. These are:

- **Fact:** The result of an experiment or observation. In the scientific context, facts are observations that have been repeatedly reproduced and confirmed so that they are regarded as 'true' for all practical purposes. However it is important to recognize that truth in science is never final and facts accepted today may be changed or even discarded tomorrow. This is quite different from the common usage in which a fact is often assumed to be an immutable truth.
- **Hypothesis:** A tentative explanation of a fact or set of facts that has not yet been extensively tested. It should lead to predictions that can either be verified or proved false.
- **Law:** A verified hypothesis about one specific aspect of the Universe that describes what it does under a particular set of circumstances. An example from physics is Ohm's law, which describes the behavior of electrical currents in certain types of materials. A physical law is usually expressed in a mathematical equation that relates two or more quantities, as in Newton's law of gravity that relates gravitational force, masses, and distances.
- **Theory:** A conceptual framework that explains existing facts and correctly predicts the results of new observations. It can include facts, laws, and verified hypotheses, and in science it exists on a higher plane than any of these. This is

doi:10.1088/978-1-6817-4100-0ch1

undoubtedly the greatest difference in terminology from common usage, where one often hears that something is 'only a theory and not a fact'.

With this background, it is possible to describe the scientific method as a list of procedures to be followed:

(1) **Observe** some aspect of the Universe.
(2) **Construct** a hypothesis that is consistent with your observations.
(3) **Use** the hypothesis to make predictions.
(4) **Test** those predictions by making further observations or experiments.
(5) **Modify** your hypothesis in the light of your new results.

After repeated application of these procedures, the hypothesis can be considered to be verified and it may be possible to incorporate it into a scientific theory. Some characteristics of a successful scientific theory are:

- **Generality:** The ideal theory should explain as many facts as possible. A theory may at times serve as a fact explained by a more general theory having a greater range of applicability. An excellent example that will be discussed further is Kepler's laws of planetary motion, which were shown by Newton to be a direct consequence of his theory of universal gravitation. In turn, Newton's theory itself is an approximation to Einstein's general theory of relativity in cases where gravitational forces are weak.
- **Testability:** A theory should correctly predict new facts or observations that can be made. This principle has also been called *falsifiability*. A theory with a limited range of applicability is at times called a model. It is especially important to guard against unwarranted extensions of a theory to areas where it has not yet been fully tested.
- **Beauty:** The simplest theory that can adequately account for a given set of facts will be preferred. In philosophy, this is known as *Occam's razor*.

It is also worthwhile to discuss the limits of the scientific method. Science seeks natural explanations for natural phenomena. At some point, current scientific knowledge becomes insufficient to provide such explanations. Scientists do not then stop trying to understand the phenomenon, but the 'explanations' typically become increasingly speculative and enter the realm of *metaphysics*, a branch of philosophy. Further work may ultimately reveal the 'true' theory, and the history of cosmology is full of examples of this type of evolution.

1.2 Some mathematics

One of the more remarkable aspects of the study of cosmology is that the subject deals with both the largest and smallest objects in the physical Universe. Because of this, it is important to review *scientific notation* which provides a compact notation for numbers by expressing them in terms of powers of ten. Consider the quantity 1.2×10^3 written in this scheme. The power of plus 3 implies that the decimal point is to be moved three places to the right to recover the number 1200. Similarly, small numbers are represented by negative powers. For example, the quantity 1.2×10^{-4}

implies that the decimal point is to be moved four places to the left, yielding the number 0.00012. Note that the extra places were filled by zeros in both these examples. However numbers may be given with a higher degree of precision, as in the case of the number 1.2567×10^4 which translates to 12 567.

Numbers in science are associated with their corresponding *units*. We will be using the *metric system* or **MKS** system in which the unit of length is the meter (m), the unit of mass is the kilogram (kg), and the unit of time is the second (s). The use of the kilogram illustrates another way to achieve a more compact notation, by the use of a prefix (in this case 'kilo') in front of another unit (in this case the gram). Some of the most-used prefixes and their translations are:

kilo (k)	10^3	milli (m)	10^{-3}
Mega (M)	10^6	micro (m)	10^{-6}
Giga (G)	10^9	nano (n)	10^{-9}
Tera (T)	10^{12}	pico (p)	10^{-12}
Peta (P)	10^{15}	femto (f)	10^{-15}
Exa (E)	10^{18}	atto (a)	10^{-18}
Zetta (Z)	10^{21}	zepto (z)	10^{-21}
Yotta (Y)	10^{24}	yocto (y)	10^{-24}

So 1 kilogram (kg) is equal to 1000 grams (g). Note that each prefix in this table is separated from the following one by a factor of 10^3, or three *orders of magnitude* (factors of ten). Also, large units are associated with capitalized prefixes (with the sole exception of kilo). Finally, one commonly used prefix that does not appear is 'centi' (c), corresponding to 10^{-2} as in the centimeter (cm) or 1/100 m.

A nice example of the concept of orders of magnitude as applied to a small fraction of the length scales we will encounter in cosmology can be found by following this link: powers of ten.

Elementary Cosmology: From Aristotle's Universe
to the Big Bang and Beyond

James J Kolata

Chapter 2

Early astronomy

Every culture has its own cosmology but we will begin with the Western tradition and the early Greek philosophers in particular, starting with Aristotle (ca 384–322 BC). Although he did make use of concepts similar to the scientific method, particularly in debates with his teacher Plato (ca 428–347 BC) on the objective reality of the material world, Aristotle's cosmology emphasized more abstract ideas such as the 'natural state or place' of an object and the concept of 'perfection'. The Universe, or cosmos, was divided into two regions, the Earth and the heavens. Substances in the imperfect always-changing earthly realm were composed of the four elements of earth, air, fire, and water. The 'natural' place of earth, the heaviest element, was at the center of the Universe, with water, air, and then fire arranged in spherical shells around it. The heavens, however, were the realm of perfection and they were composed of an entirely different element, the unchanging *ether*. Heavenly bodies (the Moon, Mercury, Venus, the Sun, Mars, Jupiter, and Saturn in that order) were embedded in concentric, touching spheres of ether, followed by the sphere of the fixed stars (see figure 2.1). Outside of this was the realm of the *prime mover* who generated all the motion of the planets by uniformly rotating the sphere of the stars. Aristotle's Universe was therefore finite (although of unknown size) and characterized by uniform circular motion based on his conception of the circle and sphere as perfect geometrical figures. This idea dominated Western cosmology until the 17th century AD, which ultimately proved to be unfortunate since it was obvious even to later Greek philosophers that the planets did not move in perfect circles at constant speeds.

The next Greek philosopher of note was Aristarchus of Samos (ca 310–230 BC), who questioned Aristotle's geocentric Universe and introduced a heliocentric cosmology in which the Earth rotated about the Sun. He also proposed that the 'fixed' stars were extremely far from the Earth, which accounted for the fact that they did not appear to move relative to each other as the Earth moved. (This apparent motion, known as *stellar parallax*, will be discussed again later on.)

doi:10.1088/978-1-6817-4100-0ch2   2-1

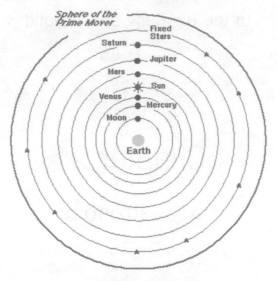

Figure 2.1. A diagram of Aristotle's Universe. Image courtesy of NASA/GFSC.

Although it conforms more closely to modern ideas, this theory did not achieve acceptance among Aristarchus' contemporaries because of Aristotle's profound reputation, the fact that it was more complicated and therefore less 'beautiful' than a geocentric Universe, and because the motion of the Earth could not be directly perceived. Aristarchus was somewhat more successful with his hypotheses concerning the relative size and distance of the Moon and the Sun. He invented the *method of lunar dichotomy*, which depends on the fact that the distance to the Sun is much larger than to the Moon. In this case, when the Moon is in its half-full (dichotomous) phase, the Sun, Moon, and Earth form a giant right triangle. If the angle to the Sun can be determined, then the relative distance to the Sun is obtained from the theorem of Pythagoras (ca 560–480 BC). Aristarchus estimated that this angle was about 87° and therefore that the Sun's distance was 18–20 times further than that to the Moon. Furthermore, he noted that the Sun and Moon subtend almost exactly the same angle on the sky (as is most clearly apparent during a solar eclipse) so that the Sun must be about 20 times larger than the Moon. The method was perfect, but unfortunately the resolution of naked-eye measurements combined with atmospheric distortion make it impossible to measure angles to an accuracy better than 2–3°. After the invention of the telescope, it was shown that the corresponding angle is approximately 89.8° and therefore that the Sun's distance is ~400 times that of the Moon. The technique ultimately gave the most accurate measurement of this ratio until well into the 18th century.

Further advances in cosmology were made by Eratosthenes of Cyrene (ca 284–192 BC). It was understood since before the time of Aristotle that the Earth was approximately spherical, based on evidence such as its shadow on the Moon during lunar eclipses and the vanishing of ships below the horizon. Eratosthenes determined the radius of the Earth to a high degree of accuracy by observing the lengths of shadows on the longest day of the year at Syene (modern-day Aswan) and Alexandria in Egypt. His original work has been lost so exact details of the method are unknown. However, it can be reconstructed by noting that the Sun is directly overhead at noon on the summer solstice (longest day of the year) at Syene, which is directly on the Tropic of Cancer. As a result, a vertical stick will cast no shadow and sunlight will cover the entire bottom of a deep well. On the other hand, Alexandria is (nearly) directly North of Syene and the Sun is never directly overhead there. The minimum-length shadow of a vertical tower in Alexandria subtends an angle of about 7°. Although trigonometry was not yet invented at the time of this measurement, the Greeks were masters of proportion and it is easy to show that the ratio of the length of the shadow of the tower to the height of the tower itself is the same as that of the distance from Syene to Alexandria to the radius of the Earth. In this way, Eratosthenes determined the radius (6400 km) to within a few percent (although there is some controversy about the actual accuracy of the measurement). Remarkably, he was also able to determine the size of and distance to the Moon. Based on Aristarchus' statement that the distance to the Sun is much greater than that to the Moon, Eratosthenes noted that the diameter of the shadow of the Earth at the Moon's distance is therefore very nearly equal to the diameter of the Earth itself, which he had just determined. By comparing the time for the shadow to completely obscure the Moon during an eclipse to the time from the beginning of the eclipse to the emergence of the first portion of the Moon from the shadow, he found that the diameter of the Earth is about 3.5 times that of the Moon. Furthermore, using the apparent size of the Moon on the sky and the fact that the angular sizes of the Moon and Sun on the sky are equal, he determined that the distance to the Moon is about 60 times the radius of the Earth, which is again within a few percent of modern values. Finally, again using Aristarchus' result that the distance to the Sun is 18–20 times that of the distance to the Moon, he found an Earth–Sun distance of about 4×10^5 km. The modern value of 1.5×10^8 km is 400 times greater, as discussed above. Nevertheless, the size of Eratosthenes' cosmos was clearly very much greater than any terrestrial distance.

The final Greek philosopher we will discuss is Claudius Ptolemy (ca 100–170 AD), who constructed a cosmology based on Aristotle's work that lasted for nearly 1500 years. Evidence had been building since at least the time of Eratosthenes that the planets did not move at constant speed in perfect circles but instead sped up, slowed down, and at times even moved backward! Also, anyone can see that their brightness changes over time, which could only happen if either their *intrinsic* brightness or their distance from Earth varied. Neither possibility is allowed in Aristotle's cosmology, in which the planets are part of the unchanging heavens and move in circular orbits around the Earth. Ptolemy devised an ingenious scheme that did a much better job of accounting for the observed planetary motions while retaining many of Aristotle's ideas (see figure 2.2). The price he paid was to introduce far more complexity into his theory by adding three geometrical constructs while retaining the basic idea of uniform circular motion. The first of these constructs was the *epicycle*: each planet moved at uniform speed on a small circle (the epicycle) centered on the circumference

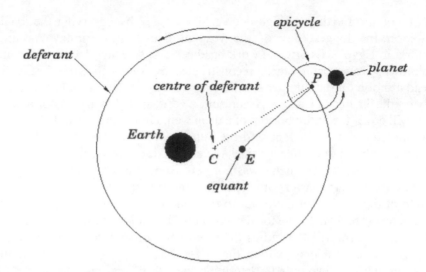

Figure 2.2. Ptolemy's scheme for planetary motion. Reproduced by permission of Professor Fitzpatrick, University of Texas at Austin.

of a larger circle centered on the Earth. By properly adjusting the speeds of rotation around the smaller and larger circles, it was possible to account for the apparent speeding up or slowing down or (at times) backward (*retrograde*) motion of a given planet. However, this change alone was insufficient to fully account for the increasingly precise measurements of planetary motion, so Ptolemy introduced the *eccentric*: the center of the larger circle was displaced by some amount from the Earth. This violated Aristotle's principle that the Earth was the center of the Universe, but only slightly—the rule was bent, not broken. Finally, even these two constructs were not enough so he added the *equant*: the center of the circle around which the epicycle rotated was also displaced from the Earth, by a different amount and even in a different direction than the eccentric. The resulting theory could account for the observed motions of the planets to an accuracy that was compatible with naked-eye measurements, which was part of the reason why it lasted so long despite its lack of beauty. It also preserved to some extent Aristotle's abstract notions of natural tendencies and the perfection of the heavens, which ultimately figured in theological doctrine.

It was only with the invention of the telescope in the early part of the 17th century that more precise observations revealing the deficiencies of Ptolemy's Universe could be made. In the intervening years, especially after the fall of Rome, the centers of cosmological research moved to China and the Islamic world. Muslim astronomers in particular, although they also worked in the Ptolemaic tradition, made major advances that eventually found their way back to the West. However, even before the invention of the telescope two people, Nicolaus Copernicus (1473–1543) and Tycho Brahe (1546–1601), began to question the Ptolemaic system. Copernicus was a Polish mathematician and astronomer who became convinced of problems with this system both by his own astronomical observations and through reading of the surviving Greek and Latin texts (including Aristarchus' work on heliocentrism).

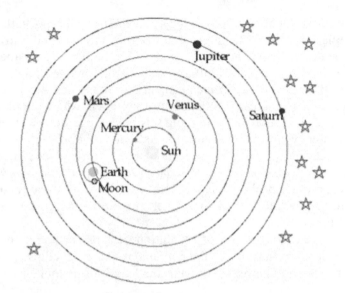

Figure 2.3. The Copernican system. Image courtesy of NASA/GFSC.

He was especially bothered by the complexities of the epicycle/eccentric/equant constructions and felt that much of this complexity would be eliminated in a heliocentric system. The publication of his book *De Revolutionibus Orbium Coelestium* (*On the Revolutions of the Celestial Spheres*) in 1543 was a major event in the history of science, initiating the Copernican Revolution. It was also highly controversial since it overthrew a cosmology that had been in place for 1500 years, but more importantly because it removed the Earth from its 'natural' place at the center of the Universe (see figure 2.3). This had profound theological implications that ultimately led to attempts to suppress the theory 60 years later.

Tycho Brahe was a Danish astronomer and nobleman who was given funding by the king of Denmark to set up an astronomical observatory on the island of Hven. His Uraniborg observatory, opened in 1576, was famous for the large astronomical instruments that he used to make measurements of the positions of the stars and planets with unprecedented accuracy. (These were still naked-eye observations since the telescope was not invented until a few years after his death.) However, his first major discovery, in 1572, was a 'new star' that suddenly appeared and then disappeared in the heavens. We now know that this was actually a supernova explosion of a star, and a remnant of that event can still be seen with a telescope. Brahe's work *De Nova Stella* (*On the New Star*) convincingly showed that this object refuted Aristotle's belief in a perfect, unchanging heavenly realm. Brahe also realized the benefits of a heliocentric model of the Universe as proposed by Copernicus. However, his preferred *Tychonic* system was a combination of geocentric and heliocentric theories in which the planets revolve around the Sun but the Earth remains at the center of the Universe. It achieved some acceptance for a while because it did not disrupt the favored natural place of the Earth, but eventually it was decisively excluded by further observations.

Modern scientific theories of cosmology had their beginnings with the work of Johannes Kepler (1571–1630), an Austrian mathematician and astronomer who tried to understand the highly accurate results of Tycho Brahe on planetary motion. After 20 years of work, Kepler eventually realized that, while Ptolemy's model did not agree with these new data, a much simpler and thus more beautiful model that does account for them emerges in a heliocentric system in which the planetary orbits are elliptical instead of circular. In making this bold suggestion, Kepler abandoned Aristotle's cosmology entirely in favor of a 'data driven' approach, rather than one based on abstract concepts such as 'perfection' and 'natural place'. He expressed his results in what have come to be called *Kepler's laws of planetary motion*:

(1) The orbit of each planet is an ellipse with the Sun at one focus.
(2) An imaginary line joining the planet to the Sun sweeps out equal areas in equal times.
(3) The square of the period of revolution of a planet about the Sun (T^2) is proportional to the cube of the *semi-major axis* of its ellipse (a^3). (The semi-major axis of an ellipse is one-half the length of its longest axis.)

It happens that the orbits of most planets differ from a circle by only a very small amount, so the first law could only be deduced from very accurate measurements. This is typical of the evolution of cosmology as a science: better instruments often lead to better theories. The second law implies that the speed of a planet as it revolves around the Sun is not uniform. When it is closer to the Sun the planet must move faster than when it is further away to satisfy the 'equal-areas' hypothesis. The third law implies that the time to complete a full revolution (the planetary year) is greater for planets further from the Sun and in a very systematic way. It states that T^2/a^3 for each planet is equal to a single constant, so once the orbit of a planet is known the length of its year can be computed and vice versa. This provides a simple explanation for the retrograde motion that so bedeviled Ptolemy. Consider as an example the planet Mars. It is further from the Sun than the Earth, which therefore at times 'catches up with' and passes Mars as they both travel in their orbits. The result is that Mars appears to travel backward for a time (figure 2.4).

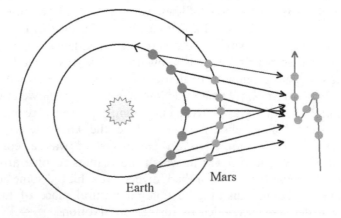

Figure 2.4. Retrograde motion. Reproduced by permission of Professor Pogge, Ohio State University.

Kepler's laws are *empirical*, i.e. they were derived from the underlying data with no formal theoretical foundation. This would be supplied, thirty years after Kepler's death, by Isaac Newton in his theory of universal gravitation.

The next important person in the history of cosmology was the Italian scientist Galileo Galilei (1564–1642). In addition to formulating the laws of falling bodies and verifying them with careful measurements, he was the first to apply the recently invented telescope to the study of the Universe. As a result, he made many important discoveries:

(1) Thousands upon thousands of stars that are invisible to the naked eye.
(2) The nature of the Milky Way.
(3) Spots on the surface of the Sun.
(4) Lunar mountains and craters.
(5) The phases of Venus.
(6) The rings of Saturn.
(7) The four (Galilean) moons of Jupiter.
(8) *Resolved* images of the planets.

The observation of spots on the Sun and mountains and craters on the Moon showed that Aristotle's heavenly realm was neither perfect nor all that different from the Earth. It turned out that the phases of Venus could only be explained in a heliocentric solar system. The motions of Jupiter's moons provided an example of a 'miniature solar system' that could be compared with the motions of the planets around the Sun. The most important of these discoveries for the eventual evolution of scientific cosmology was the fact that the Milky Way is made of an enormous number of stars. The implications of this will be discussed in more detail later. However, the firm acceptance of Copernicus' views in Galileo's *Dialogue Concerning the Two Chief World Systems* (1632) precipitated an attack on heliocentrism as being contrary to scripture. In his defense Galileo argued that scripture must at times be understood in a figurative rather than a literal sense, but this view was not accepted with the result that he was censured by the Roman Inquisition in 1633 and 'recanted' his views. Heliocentrism ultimately became the accepted theory, but Galileo's censure was not lifted until almost 360 years later.

The final historical person who shaped the leading-edge theory of the Universe at the end of the 18th century was Isaac Newton (1642–1727). He is familiar to those who have taken high-school physics for his three laws of motion, which encapsulated and extended Galileo's observations concerning falling bodies. The second law ($F = ma$) is a mathematical expression for the *acceleration* (or change of the state of motion) of an object of mass m when subjected to a force F. From the viewpoint of cosmology, however, Newton's universal law of gravitation is more important. The apocryphal story of Newton and the apple could have some truth, in that it encouraged him to speculate on the nature of the force that causes all bodies to fall. If this *gravitational force* acts on an apple on the top of a tree, why could it not also act on the Moon and other astronomical bodies? How far does gravity reach? A simple *thought experiment* may give some impression of his thought processes in trying to answer this question. When you throw an object like an apple horizontally, it falls to the ground. The faster you throw it, the further it goes before striking the ground. Newton (and Galileo before him) noticed that the rate of fall of an object does not depend on its mass (neglecting air friction of course), so you can replace the apple by a cannon ball and your arm by a cannon so that the falling object can be given an even greater velocity. In order to avoid collisions with other objects, suppose we fire the cannon ball horizontally from the top of a high mountain. Obviously, it will travel much further before hitting the ground. One reason is that the ground falls away from the top of the mountain. But remember that the Earth is approximately a sphere, so if the cannonball travels fast enough the ground will continue to fall away from it and it could in principle circle the entire Earth. (This is exactly what happens to a satellite in orbit, which is always falling toward the Earth but ends up going around it instead.) Now, Newton knew the distance to the Moon and the time it takes to go around the Earth, so he could figure out the strength of the gravitational force at that distance. This led to his universal law of gravitation, which states that the force of gravity between two objects is proportional to the product of their masses and inversely proportional to the distance between them. In mathematical terms, this law is given by the following equation: $F_{grav} = Gm_1m_2/R^2$. There are several important things to note about this expression. First of all, Newton's second law applied to falling objects has mass m_1 on one side of the equation and F_{grav} on the other side, so m_1 cancels out and objects fall with the same acceleration independent of their mass as mentioned above. The masses of the apple, the cannonball, or the Moon do not enter into the discussion at all! Only the mass of the Earth (m_2 in this equation) counts, but that is the same for all three objects falling towards (or around) the Earth. The second thing to notice is the distance R. Since the Earth is a large object, what should we take for this distance? In the end, Newton invented calculus to show that the Earth, apple, cannonball, and Moon all behave as if all their mass was concentrated at a single point. (In the case of a sphere that point is at its center.) Next, the *proportionality constant G*, known as Newton's gravitational constant, turns out to be a very small number. For this reason, it was not measured until about 70 years after Newton's death. This relative 'weakness' of gravity will play an important role in later discussions of modern-day cosmology. Finally, the law is *universal* in the sense that it applies to all gravitating bodies. For example, the motion of the planets around the Sun and the moons of Jupiter around that planet can be computed with the same equation. All that changes is the mass of the *central object* and the distances involved. As already mentioned, Newton went further and also derived Kepler's three laws from his own gravity equation, yielding a consistent account of planetary motion that was based on theory. What he did not do is explain exactly how the force of gravity was transmitted. Until Newton's work, forces were assumed to be pushes or pulls, but nothing can be observed pulling the planets toward the Sun. This 'spooky' concept of action at a distance remained controversial until the beginning of the 20th century.

Chapter 3

Nebulae

By the end of the 18th century, astronomers had a relatively good idea of the size of the Solar System. The Earth–Sun distance had been measured with a fair amount of accuracy by applying Aristarchus' method to telescopic observations. Newton's theory of gravity could then be used to determine the distance to all the planets from their known years. At this time Saturn was the furthest known planet and its distance from the Sun turned out to be a fantastic 1.4 billion kilometers (1.4×10^9 km)! However, the distance to even the nearest star was beyond the measurement capability of even the best instruments of the time. Nevertheless, observations and speculations from this period concerning the astronomical objects known as *nebulae* ultimately were important in revealing the immense scale of the visible Universe. Nebulae are fuzzy or cloudy regions in the sky (when observed with the naked eye) that do not at all resemble stars. The largest of these is the Milky Way, which was shown by Galileo to be composed of a vast number of stars. However, modern astronomers have classified four different types of nebulae:

1. Planetary nebulae; e.g. Ring Nebula (M57)
2. Diffuse nebulae
   a. Gaseous, glowing; e.g. Crab Nebula (M1/NGC1952)
   b. Dusty, absorbing; e.g. Horsehead Nebula (Barnard 33)
3. Star clusters
   a. Open clusters; e.g. NGC3293
   b. Globular clusters; e.g. NGC104/47 Tucanae
4. Extragalactic nebulae
   a. Spirals; e.g. Andromeda (M31/NGC224), Whirlpool (M51/NGC5194)
   b. Ellipticals, e.g. M87
   c. Irregulars, e.g. Small Magellanic Cloud

doi:10.1088/978-1-6817-4100-0ch3

The letters in the examples given above refer to different catalogs of these objects. For example, M57 is the 57th object in the Messier Catalog published in 1771. It originally contained 103 objects, later expanded to 110 by other astronomers. The New General Catalog (NGC), published in 1881, was an extended version of a catalog due to John Herschel, an English astronomer, taken from his observations and those of his father William. It currently contains more than 7800 objects.

Planetary nebulae, such as the Ring Nebula, are now known to result from *nova events*, in which stars at a certain stage in their evolution eject their atmospheres at high velocity leaving a central star surrounded by a glowing ring that is actually a spherical shell of hot gas. Diffuse nebulae are often clouds of gas or dust that absorb light, or emit it if they are heated by nearby hot stars. However, the Crab Nebula M1 is actually the remnant of a *supernova explosion* of a star, recorded by Chinese astronomers in 1054 AD. The explosion completely disrupted the star, but left behind an expanding and glowing cloud of gas within which a *neutron star* has been observed. Open and globular clusters are associations of various numbers of stars in our own Galaxy. Finally, the three different types of *extragalactic nebulae* are now known to be associations of a truly astronomical number of stars that are located outside our own Milky Way Galaxy. For example, the Andromeda Galaxy has been determined to contain about one trillion (1×10^{12}) stars!

None of this *nebular taxonomy* was known in the 18th century. However, the English astronomer Thomas Wright speculated in his *An Original Theory or New Hypothesis of the Universe* published in 1750 that the Milky Way was actually a vast flattened disc of stars, containing the entire Solar System as only a nearly insignificant part of the whole. Furthermore, he suggested that the spiral nebulae were each *island Universes* in their own right, similar to the Milky Way but much more distant. These ideas were picked up and expanded upon by the German philosopher Immanuel Kant in his *Universal Natural History and Theory of the Heavens* (1755) and also by the Swiss astronomer Johann Lambert in 1761. However, these hypotheses could not be verified until the development of methods to measure intergalactic distances in the early part of the 20th century.

Elementary Cosmology: From Aristotle's Universe
to the Big Bang and Beyond

James J Kolata

Chapter 4

Cosmic distances

4.1 The cosmic distance ladder

4.1.1 The parallax view

The early Greek astronomers had no way of estimating the distance to the stars and assumed that they were all at the same distance beyond the most-distant planet. In the Ptolemaic model, the most highly developed Greek cosmology (which remained the 'standard' cosmology until the time of Copernicus), that planet was Saturn at a distance of about 75 million miles. This distance estimate was based on the idea that the planets were attached to tightly nested spheres, with no empty space between the outer edge of one planet and the inner edge of the neighboring planet. Using Aristarchus' estimate of the Earth–Sun distance plus estimates of the sizes of the planets, Ptolemy arrived at a planetary system that was far too small (by about a factor of 10) compared with modern values, but still huge compared with the distance scales of the time.

With the acceptance of the idea of a moving Earth orbiting the Sun, it became possible to reconsider the notion that the stars are all at the same distance from the center of the Universe (the Sun in the Copernican model). In any moving-Earth model, the movement of the Earth should lead to an apparent annual motion of the stars known as the *stellar parallax* as the Earth moves around the Sun. Parallax is simply the change in the apparent position of a nearby object relative to a more-distant one due to a change in the position of the observer. It can easily be demonstrated by extending your arm and looking at your thumb first with one eye and then with the other. The apparent position of your thumb moves with respect to a distant wall and the motion is greater if your thumb is closer, as you can readily see. If you know the distance between your eyes and measure the angle corresponding to this apparent shift, simple trigonometry lets you compute the distance to your thumb.

In fact, parallax shifts are automatically interpreted by your brain to construct *depth perception*.

Now if the Earth moves and parallax shifts are *not* observed from one side of its orbit to another, then either the stars are all at exactly the same distance from the Sun, or they are so far away that the shifts are too small to observe. This concept was known to the ancient Greek philosophers and the apparent lack of stellar parallax was one of the arguments used to place a non-moving Earth at the center of the Universe. Aristarchus' hypothesis that the stars might be so immensely far away that the parallax shift would be unobservable was rejected as being too improbable. In fact, stellar parallax was finally observed around 1838 by Bessel, Struve, and Henderson. The largest known stellar parallax, 0.76 s of arc, was measured by Henderson for the star Alpha Centauri. One second of arc is 1/3600 of a degree, an angle so small that it can only be measured using a large modern telescope. By convention the parallax angle is taken to be $\frac{1}{2}$ the angular difference from one side of the Earth's orbit to the other, so the Earth–Sun distance forms the base of a very large triangle (see figure 4.1). Trigonometry then gives the distance to Alpha Centauri as about 270 000 times the Earth–Sun distance, or 25 million million miles $(4.0 \times 10^{13}$ km)! Since this is the closest star to us, the scale of the Universe is clearly immense. Even more remarkable, only about one hundred thousand of all the known stars are close enough that their parallax can be measured with even the most modern instruments such as the Hubble Space Telescope (HST). Because of the very large distances implied by these measurements, astronomical distances are often given in *parsecs*. One parsec is the distance at which a star would have a parallax of 1 s of arc, about 3.1×10^{13} km. Note that the further away a star is the smaller its parallax, p, so the distance is proportional to $1/p$. Another commonly used distance unit is the *light year*. One light year is the distance that light, traveling at 3×10^5 km s^{-1}, covers in one year. On this scale, the distance to Alpha Centauri is 4.3 light years and one parsec is the equivalent of about 3.3 light years.

On a historical note, Edmund Halley (of comet fame) thought he had observed stellar parallax in 1718. However, what he actually discovered was *stellar aberration*, which occurs because light takes a finite time to travel the length of a telescope (see figure 4.2). During that time the Earth is moving around the Sun, so the apparent angle from which starlight comes is shifted by an amount depending on the ratio of the speed of the Earth to the speed of light (denoted by c). While that is a very small number, it actually causes a deflection of about 40 s of arc, fifty times the parallax

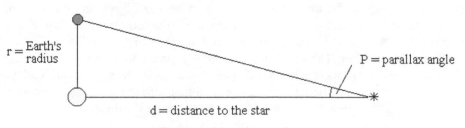

Figure 4.1. Measuring parallax.

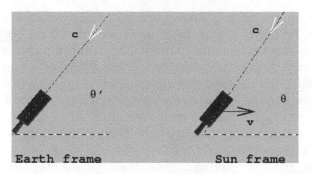

Figure 4.2. Illustration of stellar aberration. In a frame at rest relative to the Sun, the telescope moves at speed v. The light ray from a distant star shown will exit the eyepiece of the telescope at its center since the telescope moves while the light propagates down its length. As a result, the observed angle is different from that which would have been observed if the Earth were at rest.

angle of the nearest star! This effect has to be corrected for in measuring stellar parallax.

4.1.2 The color of starlight

If you look carefully at the stars in the night sky you will see that there are some subtle variations in their colors, which range from red to yellow to blue-white. This turns out to be an important observation, since it gives astronomers another way to measure the distance to stars. The parallax method is *direct*, since it depends only on trigonometry and the well-known Earth–Sun distance. Unfortunately, as mentioned above, it is limited to only the closest stars because of the small size of the parallax angle and *indirect* methods must be relied on to extend measurements to more distant stars. One of these has to do with the familiar notion that hot objects glow with a color that is related to their temperature. We speak, for example, of red-hot and white-hot objects. It turns out that the precise relationship between the color of light emitted by a hot object and its temperature was only understood with the invention of *quantum mechanics* at the beginning of the 20th century. The German physicist Max Planck found that the spectrum of colors emitted by an object could only be correctly predicted if light consists of *quanta* having a specific amount of energy that is proportional to the frequency of the light ($E = hf$). Using this hypothesis he was able to show that a hot object emits a *continuous spectrum* of light of different colors, but the color that is brightest is directly related to its temperature by the very simple equation $f = bT$. Here, f is the frequency of the light, b is a constant, and T is the *absolute temperature* of the object, i.e. its temperature in Centigrade degrees above absolute zero ($-273.15°C$). This is known as Wien's displacement law. Now, your eye responds to the average of all the colors in the spectrum, but that is roughly the same as the brightest one so the perceived color shifts from red (low-frequency light) to blue (high-frequency light) as the temperature of an emitter is raised. 'White-hot' objects are at such a high temperature that their light looks white. This idea forms the basis of an instrument known as an *optical pyrometer*, used to very precisely measure the temperature of molten steel by comparing its color to that of a filament at known temperature. The method can be made even more precise by the use of a *spectrometer* that breaks up the continuous spectrum into its component colors. Astronomers can then compare the predictions of Planck's theory directly to the observed brightness of all the colors in the spectrum of light from a star to determine its temperature. While interesting, the measurement of a star's temperature does not yet give us a method to determine its distance. That comes from another prediction of the same theory. Planck deduced the fact that the total amount of energy radiated by a given area on the surface of a star per second is proportional to the fourth power of its *absolute temperature* (T^4). This is known as the Stefan–Boltzmann law. However, the brightness of a star is also related to the amount of energy it radiates per second. Therefore, if we know the temperature of a star we can deduce something about how bright it will appear to be. There are two complicating factors: (1) a star that is further away will appear to be dimmer than one that is closer to us. But it is well known that this dimming effect is simply given by the same *inverse square* law we discussed in the context of gravity. A star of a certain *absolute brightness* that is twice as far away from us as an identical star will appear to be one fourth ($1/2^2$) as bright. Turning this relationship around, astronomers can deduce the distance of a star from its apparent brightness if its absolute brightness is known. This leads to the second complicating factor: (2) the absolute brightness of a star depends on T^4, but also on its area. Larger stars are going to be brighter and this is a problem since we have no way to directly measure their size. Fortunately, it happens that the size of a star is related (although not simply) to its temperature. Blue stars are both hotter and bigger than red stars, leading to an observed pattern of the absolute brightness of stars known as the *color-magnitude diagram* (magnitude is a term used by astronomers to refer to the brightness of a star). This pattern, sometimes referred to as the Hertzsprung–Russell (HR) diagram after its creators, is simply a plot of the absolute magnitude (brightness corrected for distance) versus the color of stars. The importance of the pattern is that the absolute magnitude of a star can be deduced from its color and then used to obtain its distance from the inverse square law. It is now known to result from an understanding of stellar physics, i.e. how stars are born and evolve. However, it was originally observed and calibrated using stars whose distances were well known from parallax measurements and the method has been confirmed and extended using the most recent astronomical observations. This 'second rung' on the ladder of the distance scale, although it is indirect, can be used to measure distances to stars that are much further away (although it is necessary to correct for the reddening and dimming of light passing through gas and dust between the star and the Earth).

4.1.3 The Cepheid variables

A variation of the HR method relies on the fact that blue stars are the brightest and there is an upper limit on their absolute magnitude. The brightest star in a distant galaxy can be assumed to be at or near that upper limit, from which its distance can be estimated. This is clearly not a very precise method, so another rung on the

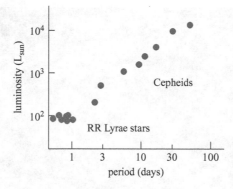

Figure 4.3. Luminosity–period diagram for variable stars. The RR-Lyrae stars all have the same luminosity and can therefore also be used to measure the distance to a galaxy. However, they are dimmer and therefore more difficult to see.

distance scale is necessary to determine the distances to other galaxies. This was provided by Henrietta Leavitt (1868–1921) through her seminal work on variable stars in the 1890s. It was known at the time that, unlike the Sun, some stars vary considerably in brightness over a period of days or weeks or months, in a precise and repeating pattern. The prototype of this category of stars was Delta Cephei, discovered in the constellation Cepheus by John Goodricke in 1784, so they have come to be called Cepheid variables. By making careful measurements of a series of photographs of the sky, Leavitt identified about 2600 of these stars (approximately $\frac{1}{2}$ the number that were known up to 1900). However, her most important contribution (for which she unfortunately received little credit during her lifetime) was the observation of the *period–luminosity relationship* (see figure 4.3). Luminosity is another term for the absolute brightness of a star, and the period is the amount of time that it takes a variable star to go from maximum to minimum brightness and then return to the maximum. In order to work out this relationship, she concentrated on stars in the *Magellanic Clouds*, now known to be small companion galaxies of the Milky Way, but then only known to be very far away compared with their size so that it could be assumed that all the stars were at the same distance in determining their luminosity. In 1912, Leavitt reported that the period of Cepheid variables was directly correlated with their maximum luminosity.

This correlation allows the distances to these variable stars to be determined from their periods, which easy to measure. In 1913 the distance to the Magellanic Clouds was measured by Hertzsprung using the HR diagram, which calibrated the method. More recent work using the HST has greatly improved this calibration. As a result, the Cepheid variable method can be used to determine the distance to very far-distant galaxies. The dimming and brightening of these stars is now known to be due to pulsations in their size. As mentioned above, the bluest stars are also both the biggest and the brightest. Like bells, larger stars pulsate at a lower frequency and therefore have longer periods.

Figure 4.4. Hubble Space Telescope image of supernova 1994D, a Type Ia supernova in galaxy NGC4526 (below the disc of the Galaxy). Image courtesy NASA/ESA, The Hubble Key Project Team, and The High-Z Supernova Search Team.

4.1.4 The supernova scale

The final rung on the distance ladder is provided by supernovae, gigantic explosions of stars that emit vast quantities of light energy which can be detected at very great distances. These catastrophic events have been studied for many years and by the late 20th century it was determined that they come in different types, distinguished by different underlying physical mechanisms of the explosion. One particular kind, the Type Ia supernova, was shown to reach a peak absolute brightness that is the same for all members of this class. The peak brightness, reached at about 20 days after the explosion, is about 5×10^9 times that of the Sun so a Type Ia supernova can outshine the entire galaxy in which it occurs (for a brief time), see figure 4.4.

However, by about 100 days after the explosion the brightness has decayed to the point that it has become difficult to see. The detailed shape of the brightness decay curve as a function of time is used to classify the event as Type Ia. Since all of these events have the same maximum absolute brightness, they can be used as *standard candles* to measure the distance to extremely distant galaxies. Note that this method assumes that the mechanism of Type Ia events is the same in these far-distant galaxies as it is in much closer galaxies. This has been the subject of some controversy, but the most recent work seems to indicate that it is a good assumption.

4.2 Spiral nebulae: are they extragalactic?

A definitive test of Thomas Wright's speculation that the spiral nebulae were *island Universes* similar to the Milky Way could not be made until the development of the Cepheid variable technique to measure very great distances in 1912. As a matter of fact, the question was still open until the American astronomer Edwin Hubble (1889–1953) used the 100-inch Hooker telescope at the Mount Wilson Observatory in California to study the Andromeda Nebula (M31). This was the largest telescope

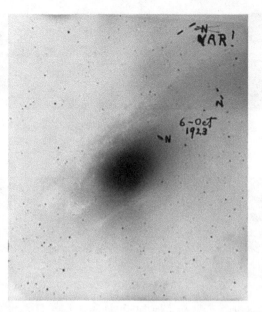

Figure 4.5. Hubble's image of the Andromeda Nebula. Courtesy Mt. Wilson Observatory.

in the world when it was completed in 1917 and Hubble used it to make photographs of M31 over a period of several nights in October of 1923 (see figure 4.5).

Upon comparing the images, he found that three new stars had appeared and then disappeared in this object. Such *novae* are now known to result from nuclear explosions on the surface of the star that generally do not completely disrupt it, so they are distinct from *supernovae*. However, after looking at more photographs Hubble realized that one of these stars was a Cepheid variable (indicated by VAR! in figure 4.5) since it brightened and dimmed in a regular fashion. He could now use Henrietta Leavitt's data to determine the absolute brightness of the star and then deduce the distance to it from its apparent brightness on the photographic plate. The result was astounding: it seemed that the star was nearly 1000 000 light years distant! Since previous work had shown that the Milky Way is only about 100 000 light-years across, M31 was clearly extragalactic. As will be discussed further, it turned out that Hubble's result was actually too small. The Andromeda Galaxy is now know to be at a distance of 2.5×10^6 light years from Earth and is the most distant object that can be seen with the naked eye. Thomas Wright's speculation was confirmed: there are an enormous number of extragalactic objects (including *elliptical* and *irregular* galaxies) and the scale of the Universe is vastly greater than even the immense size of the Milky Way.

4.3 The chemical composition of stars

In this chapter, the methods that are used to measure distances to stars and galaxies have been discussed. In subsequent chapters, it will be shown how these measurements are necessary not only to appreciate the vast scale of the Universe, but also for

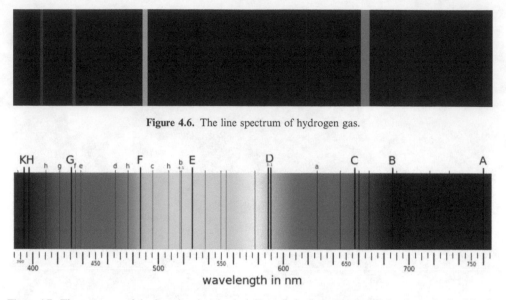

Figure 4.6. The line spectrum of hydrogen gas.

wavelength in nm

Figure 4.7. The spectrum of the Sun showing the dark Fraunhofer lines. This image has been obtained by the author from the Wikipedia website https://commons.wikimedia.org/wiki/File:Fraunhofer_lines.svg, the image is stated to be in the public domain.

the understanding of its origin and evolution. However, before moving on, it is necessary to discuss a different type of light spectrum that will become relevant in these discussions. In addition to the continuous spectrum emitted by a hot object, quantum mechanics predicts the emission of *line spectra*. These consist of several (in some cases many) narrow spectral lines at different, discrete frequencies (colors) characteristic of the element that is involved. An example is the spectrum of colors emitted by hydrogen gas. It displays four lines in the visible portion of the spectrum: the red H alpha line, an aqua line, a deep-blue line, and a violet line (there are other lines in the infrared and ultraviolet that cannot be seen with the unaided eye), see figure 4.6.

These colors can be predicted with a high degree of accuracy for all elements by the use of modern quantum theories. They come from transitions between specific energy levels within the atoms in question. Another result from quantum mechanics is that atoms absorb light at some of the same frequencies that are observed in their *emission* spectrum. This effect can be seen in the *Fraunhofer lines*, dark lines in the spectrum of the Sun caused by absorption of specific light frequencies by elements in its atmosphere, see figure 4.7.

Both emission and absorption spectra are characteristic of the atoms involved and can be used to determine the chemical composition of the Sun and other stars. However, there is one other important fact associated with line spectra, which is that the observed frequency (color) of a moving object is modified according to the *Doppler shift*. This phenomenon is familiar from your experiences of listening to sirens on moving emergency vehicles such as police cars or fire trucks (see figure 4.8).

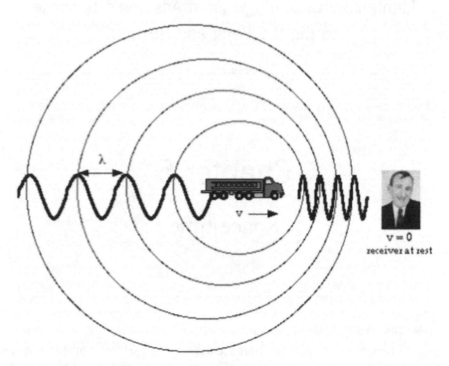

Figure 4.8. Illustration of the Doppler effect for sound waves. Image courtesy of NASA/GFSC.

As the vehicle moves toward you, the pitch of its sound is shifted upward due to the fact that successive peaks of the sound wave arrive sooner because of the motion of the vehicle. Then when it passes you the sound suddenly shifts to a lower frequency as the time between successive peaks of the sound wave is stretched out.

The analogous effect on light is the shifting to higher (bluer) frequencies of light emitted by a source moving toward you and toward lower (redder) frequencies if the source is moving away from you. It is also possible to determine the speed of the moving object from the size of the shift. The effect occurs for all types of light emission, but is difficult to observe with a broad continuous spectrum. However, line spectra are very narrow, making the Doppler effect much easier to observe.

Chapter 5

Space-time

5.1 The speed of light

Much of the progress in cosmology that occurred in the early part of the 20th century was due to an increased understanding of the nature of light. The early Greek philosopher Empedocles (490–430 BC) argued that light took a finite time to move from one place to another. Aristotle strongly disagreed, suggesting that light was not associated with movement and therefore could not be associated with a speed. He backed up this argument by noting that the motion of light could not be observed over the distance from the extreme east to the extreme west. Movement over such a large distance in an undetectable amount of time strained the imagination to the breaking point. As was the case with much of Greek thought, this argument carried the day until Galileo designed an experiment to measure the speed of light. He placed an assistant on a distant mountaintop (perhaps 2 km away) and told him to flash the light from a lantern when he saw Galileo do the same. The idea was that there would be a measurable delay due to the time it took for the light to travel from Galileo to his assistant and back again. The experiment failed of course, because the speed of light is now known to be much greater than the ability of the method to measure. Ironically, however, it was Galileo's discovery of Jupiter's moons in 1610 that eventually led to the first real measurement of this important quantity in 1676 by the Danish astronomer Ole Romer (1644–1710), who noticed that the timing of the eclipses of the Moon Io by the planet Jupiter varied with the distance between Jupiter and the Earth in a regular fashion. After careful study of this phenomenon, Romer concluded that the discrepancies could be explained if light took about 10 min to travel the Earth–Sun distance (1.5×10^8 km). This translates to a speed of $\sim 2.5 \times 10^8$ m s^{-1}, close to the modern value of 3.00×10^8 m s^{-1}.

The next important question to be answered about the nature of light concerned the medium that was 'waving'. Light was known to be a wave phenomenon ever since Thomas Young's famous *interference experiment* published in 1804. It was therefore assumed that it required a medium (called the *aether*) in the same way that waves on the ocean need water as their medium. The American physicists Michelson and Morley reasoned that light from a distant star would encounter an 'aether wind' due to the motion of the Earth and they designed an apparatus to measure it. Their idea was that the aether wind was similar to the current in a river. It turns out that a boat traveling directly across a river and back (actually traveling at an angle to compensate for the current) always arrives earlier than a similar boat traveling directly upstream and then back to the starting point, assuming that both boats move the same distance and at the same speed relative to the water. In the Michelson–Morley experiment, a light wave was divided in two using a partially silvered mirror. One of these waves traveled perpendicular to its initial direction and was then reflected backward to a detector. The other wave continued in the original direction, was reflected back by another mirror and then was reflected at 90° by the partially silvered mirror so that it entered the same detector (see figure 5.1).

If the original direction was perpendicular to the aether wind, then the second wave traveled across the stream and should arrive at the detector first. The time difference could be measured with high accuracy by the interference phenomenon. For example, if the later-arriving wave came one-half wavelength late, then its *peak* would fill the valley of the earlier wave and the result would be a dark band due to *destructive interference*. On the other hand, if it came one full wavelength late the two peaks would coincide and add together to produce a bright band. Since the speed of light is so large and its wavelength so small (300–600 nm depending on its color), the Michelson–Morley apparatus was capable of measuring very small time differences. This was important since they knew that the speed of light is 10 000 times greater than that of the Earth in its orbit so the expected effect was very small. Of course, it was not really possible to ensure that the apparatus was oriented perpendicular to the aether wind, so it was designed to rotate around its axis. Measurements were made at many different angles. In addition, data were taken at intervals of half a year so that the direction of the Earth relative to the aether presumably changed, which would increase the expected effect. Yet, despite all their care, they were never able to observe

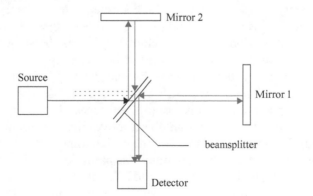

Figure 5.1. The Michelson–Morley aether wind experiment.

any effect as they reported in 1887. Subsequent attempts to repeat the experiment with improved sensitivity gave identical results—there is no aether wind and therefore no aether. This seemed to leave unresolved the question of what was actually 'waving'. However, the Scottish physicist James Clerk Maxwell had published in 1862 a set of equations that forms the basis of the science of electricity and magnetism. He showed that these two apparently different phenomena were actually two manifestations of the same effect: electromagnetism. Remarkably, his equations predict the existence of an *electromagnetic wave* which requires no 'waving' medium and moves at a speed that could be calculated from two well-measured constants relating to electricity and magnetism. This calculation gives a speed of 3×10^8 m s^{-1}, exactly equal to the measured speed of light. (This is actually the speed of light in a vacuum, as it is known to travel slower through other media.) There is no reference in the theory to either a source or an observer: the speed of light through a vacuum is a constant, independent of the motion of the source and the observer, exactly what is required to explain Michelson and Morley's result. However, this concept was very difficult for physicists of the late 19th century to accept. To understand why, consider the speed of a baseball thrown from a moving truck as measured by someone on the side of the road. Suppose the truck is moving at 60 miles per hour and the baseball is thrown in the direction of motion at 50 miles per hour relative to the truck. Then, the (stationary) observer at the side of the road will measure a speed of 110 miles per hour relative to the road—the sum of the speeds of the truck and the baseball. On the other hand, if the baseball is thrown backward its speed relative to the road is only 10 miles per hour, the difference of the two speeds. Light must be very different from baseballs in some mysterious way, since its speed is always the same: light from a moving flashlight has the same speed as if the flashlight were stationary. The mystery was only solved with the publication of Einstein's special theory of relativity in 1905.

5.2 The special theory of relativity

Albert Einstein (1879–1955) was fascinated by the idea of light and particularly the question of what someone might observe if he or she were moving at a speed near to that of light. The special theory of relativity was his first attempt to answer this question. The theory is 'special' because he restricted it to the case of an object (or person) moving at a constant speed relative to another object. In other words, the effects of accelerated motion were not included (this was the subject of his general theory of relativity). Einstein was well aware of Maxwell's equations, so he took as the first postulate of his theory the hypothesis that *the speed of light through a vacuum is constant, independent of the motion of the source and the observer*. The idea was to work out all the implications of this assumption, especially for speeds near that of light. Since speed is just the distance an object moves in a certain amount of time, it seemed clear that the answer must involve one or the other of these two quantities, or possibly both of them. Let us start with time. The nature of time had been debated by the Greek philosophers, by Isaac Newton, and by many others. However, the concept was surprisingly difficult to pin down. Newton, for example, assumed that time was an absolute quantity, flowing from past to future and unaffected by anything else. But

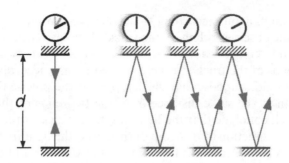

Figure 5.2. A light clock as seen by an observer moving with it (left) and one watching it move by (right). This Light-Clock image has been obtained by the authors from the Wikimedia website https://commons.wikimedia.org/wiki/File:Light-clock.png where it was made available by Michael Schmid under a CC BY SA 3.0 licence.

was this really the case? Since time is measured by clocks, we need a 'light clock' to probe what happens at very high speeds (see figure 5.2). One way to construct such a clock, at least in principle, is with a flash bulb that sends a pulse of light toward a mirror. The light pulse then bounces back to a detector and we count one 'tick' of the clock as time it takes for the flash to be detected.

Suppose we place such a clock on a truck and orient it so that the light pulse travels perpendicular to the bed of the truck. A person on the truck will see the light travel straight up and down, no matter if the truck is moving or not, and so the ticking of the clock is unaffected by the motion of the truck according to this observer. The crucial question is: *what is seen by a person standing on the side of the road*? According to this second observer, the light beam does not travel straight up and down since the truck moves while the clock is ticking. Instead, it travels toward the upper mirror on the hypotenuse of a right triangle and then back down again along a similar path, travelling a longer distance. Since the first postulate says that the speed of the light beam seen by both observers is the same, the person at the side of the road must conclude that the clock takes a longer time to tick. In other words *moving clocks run slow*. This effect has come to be called *time dilation* since it implies that intervals measured by a moving clock are stretched out. Einstein's insight was that time dilation is not specific to light clocks but rather applies to all clocks (including biological clocks). The effect has been verified many times with real clocks and, no matter how bizarre this may seem, it represents the true nature of time itself, which is therefore not an absolute quantity. We are unaware of this fundamental fact only because the effect is vanishingly small unless one travels at speeds near to that of light.

Next, Einstein considered the effect of motion on distances. Imagine a meter stick on the bed of a truck. The person on the truck measures a length of 1 m no matter if the truck is moving or not. What length does an observer at the side of the road see? In order to answer this question we need to devise a method to measure lengths of objects moving at very high speed. Einstein concluded that the only way to do this was to make measurements of both ends of the meter stick at exactly the same time. Otherwise the stick will have moved during the measurement, giving a wrong length. However, remember that we have just shown that time is not an absolute quantity, so it is necessary to determine what is meant by 'exactly the same time'. Are two events that are simultaneous according to an observer on the side of the road also simultaneous to the observer riding on the truck? The answer turns out to be no! As an illustration of this, consider a thought experiment involving two bolts of lightning that strike both ends of the truck at exactly the same time according to the observer on the side of the road. Suppose that this happens at the precise time when the two observers are both at the same distance from the two ends of the truck. Does the person on the truck conclude that the lightning bolts stuck at the same time? Since she is moving in the direction of the front of the truck, the light from the bolt striking on that end arrives before the light from the bolt at the rear and the two events are *not simultaneous* according to her. The implication of this for the length measurement discussed above turns out to be that the moving meter stick is found to be shorter than 1 m by an amount that depends on its speed. This is known as *Lorentz contraction*, after the Dutch physicist Hendrik Lorentz who first derived the equations governing the effect in an effort to explain the Michelson–Morley experiment. The contraction occurs only in the direction of motion: the width of the meter stick remains unaffected.

One conclusion from the special theory of relativity is that neither time nor space are absolute quantities. Then what is the relationship between them? In order to answer this question, let us consider a thought experiment whose results are familiar from high-school geometry. Suppose a meter stick is lying on the floor of a room with one end against a wall. Bright lights are shining from the ceiling and the other wall. Now, let us rotate the meter stick while looking at its shadows on the wall and the floor. At first, the shadow on the floor is almost equal to 1 m in length and the length of the shadow on the wall is very small. As we rotate the stick toward the vertical, the shadow on the floor becomes shorter and that on the wall becomes longer. When the stick is vertical, the shadow on the wall is 1 m in length. At all angles, however, we can recover the length of the meter stick by applying the Pythagorean theorem to the lengths of the shadows. It turns out that something similar happens in the case of special relativity. As the speed of an object increases, its length (in the direction of motion) becomes shorter while time intervals become longer. Neither space nor time are absolutes, but we can recover the *space-time extent* of an event from a Pythagorean-like theorem. While the mathematics is not quite as simple as in the case of two space dimensions, the construction shows that time is actually a fourth dimension, on an equal footing with the familiar three space dimensions, and we live in a *four-dimensional space-time*.

Finally, the most familiar result from the special theory has to do with energy. Using the equations that govern time dilation and length contraction, Einstein wrote down an expression for the *relativistic kinetic energy* of a moving object. The derivation of this quantity is just a bit more complicated than in the cases discussed above since it requires the application of some basic calculus. For those of you who are familiar with the subject, it turns out that the energy is given by a rather simple integral. As in all such cases, the evaluation of the integral results in a *constant of integration*. When Einstein evaluated this constant, it turned out to be given by the famous equation $E = mc^2$, which he interpreted as follows. In addition to the relativistic kinetic energy of motion, there is an additional energy term, equal to the mass of the object times the speed of light squared, which appears even for stationary objects. This is the *rest-mass energy* term, and it is very, very large because the square of the speed of light is a huge number. Apart from this term, the relativistic kinetic energy at a low speed, ν, approaches the value of $m\nu^2/2$ given by Newton, as it must. The equation for the rest-mass energy implies that mass can be converted to energy (and energy to mass) as was first shown with studies of nuclear reactions. We make extensive use of this fact in cosmology.

5.3 The general theory of relativity

Einstein's general theory of relativity, published in 1916, extended the study of relative motion to the case of accelerating objects. The mathematics is much more complicated in this theory, but it is possible to achieve some understanding of the results with a discussion of the nature of acceleration based on some experiences that are probably familiar to you. When you ride in an elevator that begins to accelerate upward, you feel heavier due to this acceleration. The increase in weight can actually be seen and measured if you are standing on a scale. When the elevator begins to accelerate downward, the result is a measurable decrease in your weight. If the cable is cut (and there are no brakes) you would experience *free fall* and your weight measured on the scale would go to zero. This is the exact situation for astronauts in a space station, who are always freely falling as the station orbits the Earth. If the elevator is stationary, or moving at a constant speed, your weight returns to its normal value resulting from gravitational attraction towards the center of the Earth. But now suppose that the elevator were moved to some location in space very far from any planet or star so that the force of gravity becomes negligibly small. If it now accelerates upward at exactly the

Figure 5.3. Your weight in the elevator at rest on Earth is the same as if it were accelerating upward in space at exactly the acceleration due to gravity on the surface of the Earth.

acceleration due to gravity on the surface of the Earth, your weight measured on the scale would be its normal value (see figure 5.3).

If you cannot see outside there is no way for you to distinguish the latter case from that of a stationary elevator on the surface of the Earth. In other words, gravity and acceleration are equivalent. This is a statement of Einstein's *equivalence principle* that forms the basis of the general theory, which is actually a theory of gravity.

The second thing to note is that Newton never really described how gravity, with its spooky action at a distance, actually worked. Einstein compared Newton's second law ($F = ma$) and gravity equation ($F = GmM/R^2$). The effect of gravitational attraction on an object is calculated by equating these two forces. In the process, the quantity m cancels out since it appears on both sides of the equation. This shows that *gravitational attraction is independent of the mass of the accelerating object*. Why is this? The m in $F = ma$ is called the *inertial mass* since it measures the resistance to change of motion. The m in the gravity equation is the *gravitational mass* which measures the response to gravitational forces. Apparently, these are equal; in fact, experiments have been carried out that show that they are exactly equal to better than one part in a trillion! Surprisingly, there was no compelling theoretical reason why this should be the case. With this background in mind, Einstein's insight was that the effect of the central mass in the gravity equation is to distort space-time around the object, forming a *gravity well*, as shown in figure 5.4.

The orbiting mass is simply following the path of least resistance (called the *geodesic*) in the local space-time. If there were no central object, space-time would be flat and the mass would proceed in a straight line. Warping of space-time results in an orbit. This concept solves the problem of action at a distance: the mass is simply following the geodesic in local space-time. It also explains the fact that the path depends only on the mass of the central object. The theory can be

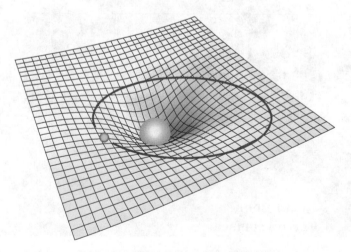

Figure 5.4. Space-time is distorted by the central mass, producing a gravity well.

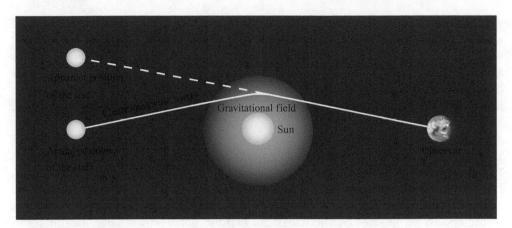

Figure 5.5. Light from a distant star bending as it passes close to the Sun.

shown to be identical to Newton's theory of gravity in the limit of weak gravitational forces.

There were three *classical* tests of general relativity, which it passed successfully. The first was a measurement, during a solar eclipse in 1919, of the bending of light from distant stars as the light passed close to the Sun (see figure 5.5).

This causes a distortion in the apparent location of the stars that was identical to the predicted value. (Actually, Newton's theory also predicts such a shift but of a smaller amount.) The second test had to do with shifts in the orbit of the planet Mercury. Such shifts can be caused by attraction to the other planets, but the value calculated with Newton's theory was too small in the amount of 43 out of 574 s of arc per century. This was a small but worrisome discrepancy, but general relativity exactly predicted it. The third test has to do with the fact that space-time is being warped, so gravity also affects the flow of time. Calculations showing that time slows down in an intense gravitational field were confirmed by the Pound–Rebka experiment in 1959. This results in *gravitational redshifts* (or blueshifts) distinct from those predicted by special relativity. Since then, the theory has successfully passed many more tests.

Once Einstein derived general relativity, he decided to apply it to the entire Universe to see what it might predict. Unfortunately, the mathematical equations associated with deformed space-time are quite complicated and could only be solved in special cases. He made the simplifying assumption that matter in the Universe is distributed uniformly in all directions and at all distances. While not the case locally, this is possibly not too far from reality when averaged over very large distances. In addition, he assumed a so-called *closed* geometry that is wrapped around on itself like the surface of a sphere, so there was no 'edge' to this Universe (but remember that we are talking here about four-dimensional space-time). The result of the calculation was a disappointment. It turned out that the predicted Universe was either expanding or contracting, but just would not stay still. This was a real problem since the favorite model of the Universe at the time, the *steady-state model*, assumed that it was always pretty much as we see it now. Expansion or contraction were difficult to accommodate in that scenario, so Einstein introduced an additional term into the general relativity equation, which he called the *cosmological constant* whose function was to keep the Universe steady. The expansion of the Universe was actually discovered less than a decade later, which led him to say that his biggest blunder was the failure to publish the original calculation. (In more recent times, however, the cosmological constant may have returned in the form of *dark energy*, which we will discuss later.)

5.4 Universal expansion

For the story of the expanding Universe, we return to Hubble's study of what he eventually knew to be distant galaxies. The story actually begins with measurements of the speeds of galaxies by the American astronomer Vesto Slipher. Beginning in 1912, he observed the Doppler shift of their spectral lines, principally the H-alpha line from hydrogen. The first result, for the Andromeda Nebula, was a very large blueshift corresponding to a speed of nearly 300 km s^{-1}! Since he could only measure the component of the speed in the direction of Earth, the true speed could actually have been even greater. This seemed much too big to be true, but he measured Doppler shifts for about 24 other galaxies by 1917 and deduced speeds up to 1000 km s^{-1}. What was even more remarkable was that 21 of the galaxies displayed redshifts, with only four blueshifts. In other words, the great majority were receding from Earth at high speed. This study was continued by Hubble and his assistant Milton Humason. By 1929 they had obtained Doppler shifts for 49 galaxies (including Slipher's original 25). Hubble's great advantage was that he could measure the distance to about half of these galaxies using Henrietta Leavitt's Cepheid variable technique. When he plotted the speed versus the distance of the Galaxy, he found a straight-line (linear) behavior. Galaxies that were twice as far away were receding from the Earth at twice the speed, on average.

This result has by now been confirmed for thousands of galaxies. The slope of the line (the rate of change of speed with distance) in called the *Hubble constant*, H. Its most recent value (as of March of 2013) is $H = 67.8 \pm 0.8$ km s^{-1} Mpc^{-1}. This means that a galaxy that is one *megaparsec* (3.3×10^6 light years) from us is receding at a speed of about 68 km s^{-1}. Most of the galaxies that do not follow this trend are part of the Local Group that are falling towards each other under the influence of their mutual gravity as, for example, the Andromeda Galaxy. In this case there is also another factor to consider. Andromeda is rotating at a speed of about 250 km s^{-1} and we happen to be viewing it edge-on. Most of the 300 km s^{-1} speed calculated from its blueshift is due to rotation, leaving only about 50 km s^{-1} for the speed of approach as Andromeda and the Milky Way fall toward each other. A related reason for differences of individual galaxy speeds from the general *Hubble flow* is that they also may be members of a cluster of galaxies. The average speed of the cluster will be given by H but there will be a spread of the speeds of individual members due to their mutual interactions, as shown in figure 5.6.

What explains this Hubble flow? Surely, there cannot be anything uniquely repulsive about the Milky Way so why do (almost) all other galaxies appear to be

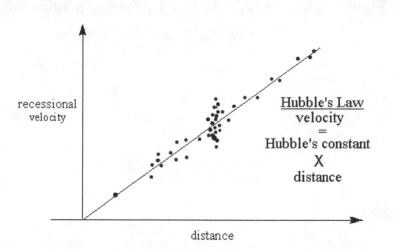

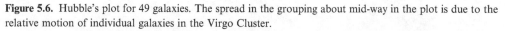

Figure 5.6. Hubble's plot for 49 galaxies. The spread in the grouping about mid-way in the plot is due to the relative motion of individual galaxies in the Virgo Cluster.

receding from us? A simple thought experiment can help here. First imagine drawing three stars in a straight line on a rubber sheet. Now stretch the sheet uniformly so that the distance between the first and second star doubles in a certain period of time. The distance between the second and third star will also double and the third star will now be four times further away from the first star. Clearly, the *recession speeds* of these stars depend on their distance in exactly the same way Hubble observed for galaxies. Next imagine that the rubber sheet forms the surface of a balloon and that there are many galaxies drawn on it. As the balloon is blown up, all the galaxies appear to be moving away from each other. The observed Hubble flow is a natural consequence of a general expansion of the entire Universe. It is also worth noting that inhabitants of any individual galaxy see all the other galaxies moving away from them. Therefore, there is no unique 'center of the Universe'. Another way of looking at this is that the center of the four-dimensional space-time expansion is in another dimension, just as the center of expansion of the balloon is outside its two-dimensional surface.

Chapter 6

The Big Bang

6.1 The structure and history of the Universe

In thinking about the large-scale structure and evolution of the Universe, it is intriguing to begin by discussing whether it could actually be infinite, eternal, and unchanging. A seemingly trivial question that sheds some light on this hypothesis is *Olbers' paradox*, first discussed by Kepler but attributed to Heinrich Olbers (1758–1840). The question is: 'why is the night sky dark?' The light from distant objects falls off as the square of the distance (the inverse square law), but the average *number* of objects at any given distance increases as the square of that distance. Every line of sight in an infinite and eternal Universe should eventually end up on a star and the entire night sky should blaze as brightly as the surface of the Sun! Olbers thought that the resolution of this paradox had to do with absorption of light by dust between us and distant stars. However, it turns out that light energy absorbed by dust grains in an eternal Universe would eventually heat them up and they would re-radiate the energy, so the paradox remains. Its resolution actually involves two concepts: the expansion of the Universe and the fact that it has a finite age. In an expanding Universe, light from distant galaxies is red-shifted by an amount that increases with distance and it is also distributed over a larger area because space has expanded since the light was emitted. Both effects imply that the intensity of visible light decreases faster than given by the inverse square law. An even more important factor, however, is the finite age of the Universe. It can be shown by a relatively simple calculation that the Universe would have to be at least 10^{16} light years in diameter for each line of sight to end on a star. We now know that the Universe is only about 10^{10} years old, so light can only have reached us from a distance that is a million times smaller than required for the night sky to blaze. The intensity of starlight is therefore at least 10^{12} times (one million squared) weaker than the light from the Sun and the night sky is very dark indeed. The paradox is resolved, but only at the expense of requiring a finite, non-eternal, and evolving Universe. In the following sections, we will see how this is exactly the kind of Universe predicted by Einstein's general theory of relativity.

6.2 The geometry of space-time

The next step in the application of the general theory of relativity to the Universe was taken by the Russian mathematician Alexandr Friedmann (1888–1925), who showed that the most natural application of the theory to the entire Universe describes an expanding Universe which emerged at a finite time in the past, unlike Einstein's static and infinitely old Universe. Friedmann's model predicted a speed of expansion that was proportional to distance. Since he reached this conclusion prior to Hubble's discovery of the expanding Universe, it was not driven by experimental observation. At first, Einstein believed this result to be erroneous, but eventually he agreed that it was in fact a possible correct description of the actual Universe. Friedmann also introduced the concept of the geometry of space-time into the discussion. This could be 'closed' like the surface of a sphere, 'open' like a saddle-shaped surface, or flat (see figure 6.1). The precise geometry depends on the amount of mass in the Universe (although this is strictly true only for a Universe described by matter and gravity alone). If there is too little matter the Universe will be open; if there is more than enough it will be closed. At the borderline between these two situations, if the amount of matter is exactly a 'critical' value, then the Universe will be flat. Since we are discussing space-time, it turns out that the geometry also has profound consequences for the evolution of the Universe. If there is too little matter (an open Universe), then gravity is not strong enough to ever slow down the galaxies and they will continue to expand forever. On the other hand, if there is more than enough matter (a closed Universe), gravity will win and the expansion will eventually come to a halt and then

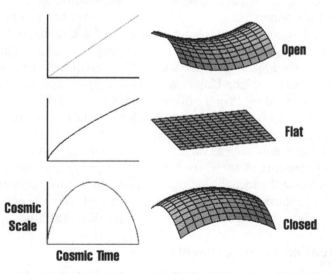

Figure 6.1. The geometry and ultimate fate of curved Universes. The 'cosmic scale' is proportional to the size of a given portion of the Universe. Image courtesy of NASA/GFSC.

turn into a contraction. The time for this to happen depends on the amount of matter present. An analogy that may be helpful is that of a rocket fired straight up from the Earth's surface. If it initially has less than the *escape velocity* (about 18 000 miles per hour), the rocket will eventually fall back to the surface of the Earth. There is a third possibility, the so-called 'flat' Universe, which stops expanding but only at infinite time; this occurs if the amount of matter is exactly the critical value. An important goal of 20th century astronomers was to determine this curvature and therefore the ultimate fate of the Universe. The question was finally answered in 1998, but in a surprising fashion that we will discuss later on.

6.3 The father of the Big Bang

An expanding Universe has as a corollary the idea that, at some time in the past, the Universe must have been smaller. One of the first to work out the consequences of this idea in the context of Einstein's theory of general relativity was Georges Lemaitre (1894–1966), a Belgian physicist. Ordained as a Roman Catholic priest in 1923, Lemaitre studied math and science at Cambridge University and eventually became a professor at the University of Louvain. Independently of Aleksandr Friedmann, he repeated Einstein's earlier calculations. By that time, however, Hubble had demonstrated that the Universe was actually expanding so the *cosmological constant* was unnecessary. Working backward from the present state of the Universe, Lemaitre realized that it must once have been much smaller and denser than it is now, with all of its matter compressed into a very small volume. This led him to propose the idea of a *primordial atom* as the original state of the Universe. A further and very important observation of his had to do with the fact that a gas heats up when it is compressed and cools when it expands. The heating can be experienced, for example, when a bicycle tire is pumped to high pressure, and cooling by expansion of a gas is used in refrigerators and air conditioners. Lemaitre therefore concluded that his primordial atom must have been very hot and he suggested that the light emitted from this hot object might still be around, but cooled by the subsequent expansion of the Universe (more about that later). This was the first explication of the *hot Big Bang model*, which has subsequently become the accepted theory of the formation of the Universe. Lemaitre also proposed that the chemical elements were produced by fission of this primordial atom. We now know that this would have led to a Universe dominated by elements near iron and that all but the very lightest elements were actually formed via fusion processes in the cores of stars or in violent stellar events such as supernova explosions. Nonetheless, the concept of the creation of elements in the early Universe was an important idea. Lemaitre's work did not achieve the recognition that it deserved, possibly because the idea that the Universe had a 'birthday' was too similar to biblical accounts of a creation event. Nevertheless, he is aptly (and only partially in jest) called the 'father of the Big Bang'.

6.4 The creation of the elements

Lemaitre's speculations were forgotten for nearly 20 years until George Gamow independently revived the concept in 1948. The problem that Gamow and his collaborators were working on when they rediscovered Lemaitre's work was the origin of the chemical elements, which they proposed (in 1946) to have occurred in the early Universe. Nowadays, we call this subject *Big Bang nucleosynthesis* (BBNS), but in 1946 the *steady-state* model was in vogue. In this model, the additional space formed by the expansion of the Universe was filled by the continuous creation of matter. Otherwise, the density of matter would be negligibly small (remember that the Universe was supposed to be infinitely old in this theory). Only a small rate of creation was necessary, but a mechanism for continuous creation was never specified. Gamow and his collaborators Ralph Alpher and Robert Herman, were familiar with the emerging discipline of *nuclear physics* and suggested that the elements could have been formed by nuclear fusion reactions in the early, hot Universe. Fusion seemed to be a realistic starting point since they knew that hydrogen and helium, the lightest elements, together make up about 99.99% of the atoms in the Universe. All other elements account for only about 0.01%. In their calculations, the process began about 1.5 min after the Big Bang, when the temperature of the Universe was about 900 million degrees, about 60 times hotter than the center of the Sun! (The time estimates given here are from more recent calculations but are close to the values they deduced). At earlier times, the temperature was just too high for even an atomic nucleus to survive. Nuclear reactions occurred at a great rate, but the nuclei were broken apart as fast as they were formed. However, at just a bit less than a billion degrees, a *deuterium* (or heavy hydrogen) nucleus can be formed by the interaction of a neutron and a proton, see

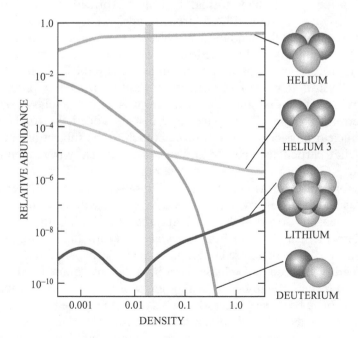

Figure 6.2. Relative abundances of the elements formed in the Big Bang plotted as a function of the density of the Universe as a fraction of the *critical density* at which the Universe would be flat. The grey bar shows the region (~4% of the critical density) where the calculation agrees with observation.

figure 6.2. This is the key reaction since all the elements heavier than hydrogen had to be formed by reactions that involve deuterium. In one such series of reactions, deuterium interacts with a proton to form helium-3, which subsequently captures a neutron to form the most common *isotope* of helium, helium-4. (Isotopes of a chemical element have the same number of protons but differ in the number of neutrons in their nucleus.) From this point on the nuclear reactions proceed very rapidly. It turns out, however, that only hydrogen, helium, and lithium were made in the Big Bang, as seen in figure 6.2 (together with the mass-7 isotope of beryllium which is radioactive and decays into lithium within a few years). This is because there are no stable isotopes with masses five or eight times the proton mass. Reactions can proceed around the mass-5 gap by fusing helium isotopes, but the nuclear physics is such that there is no good way around the mass-8 gap. Two other important effects had to be considered. First of all, the temperature of the Universe continues to fall as it expands and when about 4 min had elapsed it was so low that nuclear reactions could not occur anymore. Secondly, about 0.01 s after the Big Bang the numbers of neutrons and protons were the same due to nuclear reactions with other particles. However, about 0.1 s later when the Universe had cooled to about 30 billion degrees these reactions were not able to keep the neutron–proton balance and the number of protons began to grow at the expense of the neutrons. By the time of nucleosynthesis, the number of neutrons was only 13% of the total number of *nucleons* (neutrons and protons are collectively referred to as nucleons). Since two neutrons and two protons join to make helium-4, the most abundant isotope (other than normal hydrogen) formed in the Big Bang, the Universe at this stage is about 74% hydrogen and 26% helium, which is essentially the same as the elemental abundance observed in the Universe today.

Gamow and company had worked out this entire scenario by the early 1950s. Their most famous research paper on the subject was published on 1 April 1948 in the journal *Physical Review*. The publication date (April Fools' Day) was purely accidental but significant because Gamow, who was a bit of a 'class clown', decided to leave Herman's name off the paper and instead added that of Hans Bethe, a prominent nuclear physicist who contributed much to the subject of nucleosynthesis but little to that particular paper. In any event, it eventually became famous as the 'Alpher–Bethe–Gamow' (or alpha–beta–gamma) paper. (Today, this little stunt would probably get Gamow cited for scientific misconduct.) However, at the time the paper was widely disregarded, just as had happened with Lemaitre's work. There were several reasons for this. First of all, the theory did not account for the 0.01% of elements heavier than lithium. Now we know (and Gamow and others eventually showed) that these are either made by nuclear fusion in the cores of stars that subsequently explode and release them into the environment, or in the violent supernova explosions themselves. Although work is still continuing, nuclear astrophysicists have been very successful at accounting for the observed abundances of these other elements. It is truly remarkable that the atoms that make up ourselves, and much of the world we see around us, are literally 'stardust'. The second factor leading to skepticism about the theory had to do with the time frame for the Big Bang itself. Hubble's value for the expansion rate of the Universe led to an estimate of approximately 2 billion years for its age, significantly smaller than the ~5 billion years that had already been deduced for the age of the Earth from measurements of the decay of radioactive elements in its crust. A theory that predicts that the Earth is older than the Universe is clearly in trouble. Ultimately, this problem was resolved when it was realized that Cepheid variable stars occur in two different forms. One of these types is significantly brighter than the other and therefore easier to detect in distant galaxies and it has a different brightness–period relationship. As a result, Hubble's distance measurements (and therefore his estimate of the age of the Universe) were much too small. Finally, Gamow and his collaborators agreed with Lemaitre that the flash of light from the Big Bang should still be visible, although cooled by the expansion of the Universe, and they estimated a current temperature of roughly 5 K (centigrade degrees above absolute zero). According to Planck's quantum theory, the peak in the spectrum of this 'light' should actually occur in the microwave region, but nothing like this had ever been observed. In retrospect it is clear that instruments at the time were just too insensitive to detect these micro-waves, but this 'failure' was another significant reason why the steady-state model continued to be the preferred theory of the Universe for 15 more years.

Chapter 7

Cosmic microwave background radiation

7.1 The 'smoking gun' of the Big Bang

A profound paradigm shift in our understanding of the origin and evolution of the Universe occurred in 1964. At about that time two physicists, Arno Penzias and Robert Wilson, were working to improve the sensitivity of microwave receivers at Bell Telephone Laboratories in Holmdel, New Jersey. Their interest in this subject resulted from the fact that artificial satellites had only recently been launched. It seemed clear that long-distance communications between continents could be greatly improved by using satellites to bounce microwaves over very large distances. However, this required microwave receivers that were very much more sensitive than any that had been made up to that time, since satellites in orbit are far from the Earth's surface and microwaves bounced from them would be very weak. Penzias and Wilson produced several generations of microwave receivers, each more sensitive than the last, by reducing the *noise* generated by the electronic components in their receivers. (This noise is familiar as the hissing sound heard between stations on the radio dial, or the 'snow' on unused analog TV channels.) However, their most sensitive receiver had a noise level that they could not account for or reduce based on their knowledge of electronics. Thinking that this excess noise might somehow be generated in cities, they aimed their microwave antenna at New York City but observed no difference. Similarly, aiming their antenna at the Sun had no effect; it seemed that the noise was the same in every direction. Eventually they became aware of a paper by Robert Dicke of Princeton University who had independently arrived at the same prediction that Gamow, Alpher, and Herman had earlier made concerning microwaves from the Big Bang. Penzias and Wilson had accidently found the *cosmic microwave background* (CMB) radiation. They received the 1978 Nobel Prize in physics for this discovery. The CMB is a natural result of the Big Bang scenario but it could not be accounted for in the

doi:10.1088/978-1-6817-4100-0ch7

steady-state model, which rapidly lost favor with cosmologists. Work since 1964 has shown that the CMB displays a spectrum in exact agreement with the prediction of Planck's quantum theory for a temperature of 2.7 K. It is also very uniform in all directions. We are living inside a really, really big microwave oven that is the afterglow of the Big Bang!

7.2 Decoupling

Lemaitre suggested that what we now call the CMB was emitted by the very hot primeval atom at the time of creation. However, it turns out that this is not exactly the case because of some important properties of light. First of all, in Planck's quantum theory light comes in packets or *quanta* having an energy that is proportional to the frequency. We experience the frequency of visible light as colors, but such *electromagnetic radiation* comes in a continuum of frequencies ranging from very low (radio waves) to very high (x-rays and *gamma* radiation). In 1905, Einstein used Planck's quantum hypothesis to derive a theory of the *photoelectric effect* in which light can interact with atoms and eject electrons from them. In this theory, light behaves like a particle even though it also has wave-like properties. (This so-called *particle-wave duality* is a characteristic of all quantum theories.) We call the quantum of electromagnetic radiation a photon and the energy of each photon depends on its associated frequency according to the simple equation $E = hf$, where h is *Planck's constant*, a very small number. Since h is so small, it takes a very large number of photons to produce a reasonably bright light. The second important property of light is that it interacts very strongly with charged particles such as electrons or atomic nuclei, but far less strongly with *neutral* atoms. An example is the hydrogen gas that makes up much of the surface of the Sun. Hydrogen gas at room temperature is transparent to visible light, which interacts only weakly with its atoms. However, the temperature on the surface of the Sun is about 5500°C. Photons at this temperature have enough energy to eject electrons from hydrogen and all the H atoms on the surface of the Sun are *ionized*, that is, they have no bound electrons. The electrons and atomic nuclei roam about free of each other in a state of matter called a *plasma*. In this state, light interacts very strongly with the hydrogen plasma and the Sun's surface is opaque. Photons from below its surface take a long time to reach it, interacting very strongly along the way, and so we cannot see into the heart of the Sun. The same thing happened to the Universe during the course of its evolution. At the present time the Universe is very dilute, with only about one atom (mostly H) in each cubic centimeter of (intergalactic) space. It is also very cold (2.7 K) so that light interacts only weakly with these atoms and space is transparent. This is also true of the photons in the CMB. Now, imagine what happens as we turn the clock back to earlier times. In other words, imagine running a video of the evolution of the Universe backwards. It will become denser, so there will be more atoms to interact with in a given volume. But more importantly its temperature will rise. When the temperature of the Universe reaches approximately 3000 K, each photon in the CMB will have enough energy to ionize H atoms. At this point, the entire Universe becomes an opaque and glowing plasma. Nothing from an earlier era can be seen since the light from earlier times interacts so strongly with the plasma. According to the most recent calculations, this occurs at about 400 000 years after the Big Bang. The microwave photons we detect at the present time were emitted in that era, not at the Big Bang itself. If we now run the video forward, at 400 000 years after the Big Bang the Universe cools below 3000 K and becomes transparent as electrons and atomic nuclei combine. This process is called the *decoupling* of photons from matter. It has also been called *recombination*, although that is something of a misnomer since the electrons and atomic nuclei had never previously been combined.

7.3 How bright is the CMB?

It might be argued that the discussion in the previous paragraph omits all the photons emitted by the stars and galaxies in the Universe, and that is true. However, measurements of the brightness of the CMB have revealed a remarkable fact: the number of photons emitted by all the stars in the Universe since their formation is only 3–10% of the number of photons in the CMB! For all practical purposes, the CMB is by far the most luminous 'object' in the Universe and the effects of starlight can be completely neglected in working out the physics of the Big Bang. It also turns out that BBNS is very sensitive to the brightness of the CMB, specifically to the *photon–nucleon* ratio in the Universe. In order to match the observed abundances of all the light elements, this ratio (designated by the Greek symbol η) must be approximately 1×10^9. In other words, BBNS implies that there were approximately one billion photons in the CMB for every neutron and proton in the Universe. Since η does not change with time, that is also true of the present state of the Universe.

7.4 'Matter dominated' versus 'radiation dominated' Universes

Since there are so many CMB photons for every nucleon in the Universe, each of which has an *effective mass* given by $E = mc^2$, it might seem that they would contribute significantly to the matter–energy density of the Universe (the average amount of mass and energy in a given volume). However, as noted above, the Universe is currently very cold and it turns out that each nucleon contributes 4000 times the mass-energy of one billion CMB photons. As a result, the Universe is currently *matter dominated*. However, this was not always the case. If we again run the video of the evolution of the Universe backward, the densities of both nucleons and photons increase as the volume becomes smaller, but the mass-energy of the nucleons remains the same and their kinetic energy increases relatively slowly, while photons become much more energetic as the temperature rises. (The photon energy increases as the fourth power of the temperature according to Planck's equations.) At some point, then, the matter and radiation energy densities will become equal and this occurred at about 50 000 years after the Big Bang. Before that time, the Universe was *radiation dominated*. At the time of nucleosynthesis, the matter density was so small that it can essentially be neglected and radiation density controlled the evolution of the Universe.

Finally, the photon-to-nucleon ratio η can also be related to the matter–energy density Ω required to 'close' the Universe. By definition, if $\Omega > 1$, then the Universe is closed; if $\Omega < 1$, the Universe is open; and if $\Omega = 1$, the Universe is flat. The value $\eta = 1 \times 10^9$ corresponds to $\Omega = 0.04 \pm 0.01$. In other words, BBNS makes the remarkable prediction that the density of 'normal' matter (made of neutrons and protons) is only 4% of the density required to close the Universe. This will become important when we discuss the subject of *dark matter*.

7.5 How uniform is the CMB?

It was expected that the CMB should be very uniform since plasma filled the entire Universe just prior to photon decoupling. This turns out to be the case, but the deviations from that overall uniformity give very important information on the structure and evolution of the Universe. These non-uniformities are best discussed as deviations from the average 2.7 K temperature of the CMB photons. Many of these photons are absorbed by the Earth's atmosphere, depending upon their energy, so early measurements used balloons and rockets to fly microwave receivers above much of the atmosphere. These experiments found the first indication of a deviation at the level of a few millikelvin, or about 1/1000 of the average temperature. The temperature was higher in one direction and lower in the exact opposite direction in space. This is the so-called *dipole anisotropy*, which is now recognized to be due to the motion of the Earth relative to the CMB (see figure 7.1). Just like the Andromeda Galaxy, the Milky Way is moving through space and rotating, carrying the Solar System along with it. When we look in the direction of motion, the CMB photons are Doppler blue-shifted upward in energy, corresponding to a higher temperature. In the opposite direction, they are red-shifted downward in energy, corresponding to a lower temperature.

The next generation of experiments used satellites above essentially all of Earth's atmosphere and very sensitive microwave receivers cooled to the liquid helium temperature of about 4 K. The first of these was the Cosmic Background Explorer (COBE) experiment, launched in 1989. It observed the CMB for approximately four years and found two other types of temperature deviations (after subtracting the

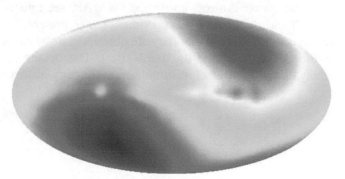

Figure 7.1. The dipole anisotropy. The temperature differences are on the order of a few millikelvin. Image courtesy NASA/WMAP Science Team.

Figure 7.2. WMAP image of the Milky Way (superimposed on the intrinsic anisotropy). Image courtesy NASA/WMAP Science Team.

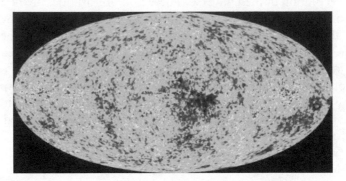

Figure 7.3. WMAP image of the intrinsic anisotropy. Image courtesy NASA/WMAP Science Team.

dipole anisotropy) at a level of about 30 µK, or 1/100 000 of the average. One of these traced the image of the Milky Way on the sky; our Galaxy is a weak microwave emitter.

The second, much more interesting and important, was a patchy and seemingly random temperature fluctuation, the *intrinsic anisotropy*, which corresponds to density deviations in the very early Universe. These density ripples are believed to be the source of the large-scale structure in the Universe today, and were further investigated by the Wilkinson Microwave Anisotropy Probe (WMAP; see figures 7.2 and 7.3) experiment (2001–10) and the PLANCK experiment (2009–present) launched by the European Space Agency. It turns out that they are now understood as due to processes predicted by quantum theory (to be discussed later). The details of the observed fluctuations contain a treasure trove of information on the early history and evolution of the Universe.

Elementary Cosmology: From Aristotle's Universe
to the Big Bang and Beyond

James J Kolata

Chapter 8

Dark matter

8.1 Dark matter defined

By definition, *dark matter* is matter that can only be detected by its mutual gravitational interaction with other objects. Its importance can best be described in the context of the matter–energy density Ω of the Universe. As has previously been mentioned, the Universe is closed if $\Omega > 1$. It turns out that all the *luminous* matter astronomers directly see with their telescopes (stars, dust, gas clouds, etc) corresponds to $\Omega_{luminous} = 0.004$, that is, less than $\frac{1}{2}$ of one percent of the amount necessary to make the Universe 'flat'. However, Fritz Zwicky suggested in 1933 that there must be a lot more matter around, based on his studies of the motion of galaxies in clusters. He found that the observed speeds of galaxies in the large Coma cluster could only be understood if the total mass of the cluster is at least 400 times greater than its luminous mass. His work was largely disregarded until about 35 years later when Vera Rubin and her collaborators began to study the *rotation curves* of galaxies. She found that spiral galaxies do not rotate the way they should if their mass were distributed according to the distribution of luminous matter. Since the majority of the luminous matter is concentrated in the galactic center, the material in the spiral arms should rotate more slowly as the distance from the galactic center increases, just as the outer planets revolve around the Sun slower than does the Earth. Instead, the observed rotational speed remains constant with distance from the galactic center (see figure 8.1). This observation can only be understood if the spiral galaxies are assumed to be embedded in huge spheres of dark matter. The total mass of a galaxy then turns out to be about 5–10 times larger than its luminous mass, so $\Omega_{galactic} = 0.02$–0.04. This is less than or equal to $\Omega_{BBNS} = 0.04$ which implies that all this matter could be made of normal protons and neutrons, referred to as *baryonic* dark matter.

As it happens, candidates for baryonic dark matter have already been detected. These are the *massive compact halo objects* (MACHOs) which have been located by

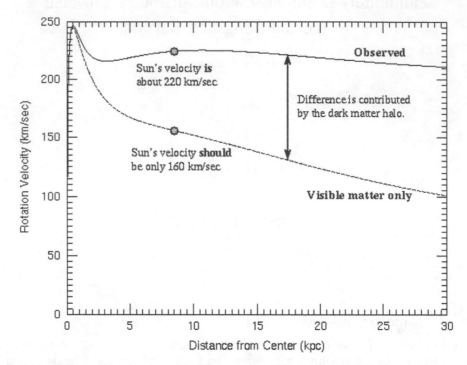

Figure 8.1. Comparison of our Sun's speed as it rotates around the center of the Milky Way with a calculation of the predicted speed if our Galaxy contained only the amount of visible matter.

gravitational microlensing. Light from a distant galaxy passing near an object in the galactic halo (the spherical cloud of non-luminous matter suggested by Vera Rubin) will be bent according to the theory of general relativity. The object in effect acts as a lens to focus the light onto a telescope on Earth if it happens to be in the correct place. Since the MACHO is moving, the result is a transient brightening of the light from the distant galaxy and much information can be derived from the precise form of the brightening curve, see figure 8.2.

The exact nature of MACHOs is unknown. There may be many types of them, such as isolated *black holes*, Jupiter-sized planets that have escaped from their planetary systems, *black dwarves* (an end stage in the evolution of stars like the Sun), or *brown dwarves* (sometimes called 'failed stars') which are gas clouds that heated up from gravitational contraction but did not have enough mass to reach a temperature high enough to ignite nuclear fusion. Other possibilities have also been suggested.

8.2 Non-baryonic dark matter

Once the idea of dark matter became more accepted, further research on Zwicky's hypothesis began, looking for example at the effect of the Great Attractor in Centaurus on the motion of the Milky Way. (The Great Attractor is mainly responsible for the dipole anisotropy of the CMB.) This work eventually reached the conclusion that Ω_{matter} is roughly 0.3 (a more precise value is 0.27 as will be

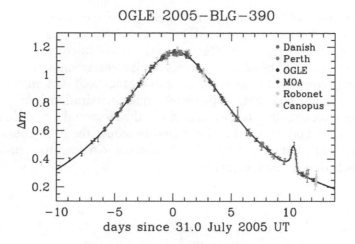

Figure 8.2. Brightening curve for a microlensing event. The 'blip' at about 10 days is due to a planet orbiting the lensing star. Reproduced with permission from Martin Dominik, University of St Andrews.

discussed later on). Though this is still not enough to close the Universe, it is far greater than the limit on baryonic mass from BBNS. It appears that most of the matter in the Universe is exotic *non-baryonic* dark matter, not made of protons and neutrons! The nature of this stuff is still under investigation. Several candidates from the realm of high-energy physics have been suggested. One of the first was *massive neutrinos*. Neutrinos are uncharged (neutral) particles, predicted in the 1930s and finally discovered in the 1950s, that interact very weakly with normal matter. Big Bang theories predict that there should be many *relic* neutrinos in the Universe, more than the number of photons in the CMB. At any one time, about twenty million of these cosmic neutrinos are zipping through your body, but because of their very weak interactions they are incredibly difficult to detect. For example, a neutrino produced in nuclear fusion reactions in the center of the Sun can travel through a light year of lead shielding without reacting. Despite this, solar neutrinos have been detected using very large and sensitive detectors. The relic neutrinos from the Big Bang have much lower energies and are even more difficult to observe, a task that is well beyond the current state of technology. If they could be detected, however, they would give important information about the history of the Big Bang. As mentioned in chapter 7, the CMB decoupled from atoms at about 400 000 years after the Big Bang. A similar thing happens for relic neutrinos, but because of their very weak interactions neutrinos decouple when the density of the Universe was much higher, at about 1s after the Big Bang. The relic neutrino background therefore carries information on the state of the Universe during the time of nucleosynthesis.

It was once thought that neutrinos had no rest mass, just like photons. Recent experiments have shown, however, that they do have a small amount of rest mass. Their exact mass is unknown, but the best current estimates give $\Omega_\nu \sim 0.001$, far too small to have any significant impact on the curvature of the Universe. Also, with such a small mass their speeds during galaxy formation would have been very high. Calculations indicate that this *hot dark matter* would completely disrupt galaxy formation so massive neutrinos do not seem to be the answer. What is needed instead is *cold dark matter*, one example being *weakly interacting massive particles* or (WIMPs). These are hypothetical (at present) weakly interacting particles with a large rest mass predicted by certain theories of high-energy physics. Experiments are currently being conducted to look for these and other cold dark matter candidates. Examples are the SuperCDMS experiment under preparation in a nickel mine in Sudbury, Ontario, and the LUX experiment in the Homestake mine in Lead, South Dakota. They are both situated underground to reduce the background from high-energy cosmic radiation reaching the Earth's surface. No definitive conclusions have yet been reached from these experiments.

Chapter 9

The standard model of cosmology

9.1 Nucleosynthesis

In his 1977 book *The First Three Minutes*, Nobel-prize-winning physicist Steven Weinberg discussed the evolution of the Universe after the Big Bang in terms of a movie whose individual frames represent its state at a given time. Since he was mainly interested in nucleosynthesis, his first frame began at 10^{-2} s after the Big Bang and each subsequent frame was at an order of magnitude greater time than the previous one. In this section we will first review this movie, which describes what is known as the *standard model of cosmology*. To begin with, however, we need to introduce some definitions (a few of which have already been encountered):

- **Nucleon:** A generic name for a neutron or proton.
- **Photon:** Einstein showed in 1905 that electromagnetic radiation (light) displayed both *particle-like* and *wave-like* behavior. (This *particle–wave duality* is a common feature of quantum mechanics, and applies in all situations.) A *particle* of light is called a photon (symbol = γ).
- **Neutrino:** A particle predicted in the 1930s and first observed in the 1950s. It is similar to an electron, except that it has no electrical charge (symbol = ν). Until recently, it was also thought to have no *rest mass* like the photon. Neutrinos interact very weakly with matter. For example, a 'typical' neutrino produced by nuclear reactions in the Sun (a *solar neutrino*) can travel through light years of lead before interacting, so solar neutrinos tell us about conditions at the center of the Sun. In the very early hot and dense phase of the Universe, however, neutrino reactions were quite common and they kept a balance between neutrons and protons.
- **Rest-mass energy:** Often referred to as just 'rest mass', this is the energy possessed by a non-moving particle simply due to its mass, computed via $E = mc^2$. The photon has zero rest mass since a photon cannot be at rest

doi:10.1088/978-1-6817-4100-0ch9

(though it can be said to have an 'effective mass' computed from its energy). Similarly, any particle that has zero rest mass (as was thought to be the case for the neutrino until very recently) *must* travel at the speed of light.

- **Electron-volt (eV):** A unit of energy equal to 1.6×10^{-19} J. This is the amount of energy gained by an electron when it is accelerated across an electrical potential difference of 1 V. An electron-volt is not a very large amount of energy. For example, a flea (mass $= 10^{-4}$ kg) jumping upward a distance of 1 m expends an amount of energy equal to 6×10^{15} eV. Particle accelerators are rated by their maximum energy in electron-volts. The most powerful accelerator at present is rated at 7×10^{12} eV (the LHC at CERN).

- **Antiparticle:** Each type of particle has associated with it a corresponding *antiparticle* which has the same physical properties except that it has the opposite electrical charge. For example, the antiparticle of the electron, the antielectron (symbol $= e^+$), has a positive charge (so it is often called the *positron*) but otherwise it looks identical to an electron. Antiparticles do not exist in great numbers in our Universe, which is something that will be discussed later, but they can be produced in nuclear reactions. However, recent results by the PAMELA and FERMI spacecraft have found positrons in Earth orbit and in violent thunderstorms.

- **Pair production:** The laws of physics allow for the conversion of energy into mass. However, it turns out that the rules require *equal* amounts of matter and antimatter to be produced in any given reaction, so that the *net* amount of matter (counting antimatter as 'negative' matter) remains equal to zero. For example, it is possible to convert a photon into an electron–antielectron pair (pair production). The photon must have a minimum energy equal to the sum of the rest-mass energies of the two particles, each of which has a rest-mass energy of 0.511 MeV in this case.

- **Annihilation:** When a particle and its antiparticle meet, both disappear in a flash of energy in a process called *annihilation*. The amount of energy produced equals the rest-mass energy of the pair (1.02 MeV in the case of an electron). It is fortunate that antimatter is not common in the Universe, since if you were to touch an 'antiperson' both of you would disappear in a blaze of energy.

- **Decoupling:** A term that refers to the situation when some type of particle ceases to interact strongly with the surrounding matter. The decoupling of photons from atoms has already been discussed.

- **Baryon number:** A term that refers to the number of heavy (i.e. high-mass) particles, such as protons and neutrons, in a reaction. It is positive for particles (protons, neutrons) and negative for antiparticles (antiprotons, antineutrons). Baryon number is *conserved* in a nuclear reaction. In other words, the number of baryons before and after the reaction are the same, although one type of baryon may be changed into another type.

- **Lepton number:** A term that refers to the number of light (i.e. low-mass) particles, such as neutrinos and electrons, in a reaction. It is positive for particles (neutrinos, electrons) and negative for antiparticles (antineutrinos, positrons). Lepton number is also conserved in a reaction.

One simple application of these definitions is to the radioactive decay of the neutron, an unstable particle that decays into a proton. The neutron is slightly heavier than the proton, by about 1 MeV in rest-mass energy, so neutrons can decay into protons. Since electrical charge is conserved in every reactions, the decay also involves a negatively charged particle, the electron. However, conservation of lepton number requires that the lepton number in the final state be zero since there were no leptons before the decay. As a result, neutron decay also results in the emission of an antineutrino (symbol $= \bar{\nu}$) so that the total lepton number after decay is still zero. This is diagrammed as n $\rightarrow p + e^- + \bar{\nu}$. The rest-mass energy of the electron is 0.511 MeV and that of the neutrino is very small, so the emitted leptons leave the interaction region with a total energy of about 0.5 MeV. It is also clear that both baryon number and lepton number are conserved, as required by the rules of the Universe. At this point, one important question that often comes up is the difference between the neutrino and its antiparticle. Both particles have zero charge, so how do they differ? It is known in fact that they do react differently, but the precise nature of the difference is still under investigation. A slightly more complicated case is the conversion of a proton into a neutron. Since the neutron is heavier than the proton, this can only occur if an incoming particle brings in enough kinetic energy to make up the mass difference. The process is diagrammed as $p + e^- \rightarrow$ n $+ \nu$. The conversion of a neutron into a proton in a similar reaction may also occur, diagrammed as n $+ e^+ \rightarrow p + \bar{\nu}$. *Inverse* reactions with neutrinos are also allowed: (n $+ \nu \rightarrow p + e^-$; $p + \bar{\nu} \rightarrow$ n $+ e^+$). Note that the identities of the particles involved are required by conservation laws for charge, lepton number, and baryon number.

9.1.1 The first frame ($t = 10^{-2}$ s)

Working backward from the present state of the Universe, its density at 10^{-2} s was about 4×10^9 times the density of water. As discussed previously, this was mostly in the form of radiation and not matter since the Universe was radiation dominated at the time. Its temperature was about $T = 10^{11}$ K, which implies that the average energy of a photon in the CMB was 10 MeV and, as now, there was only one nucleon for every 10^9 photons. Since electron pair production occurs for photon energies greater than 1.02 MeV, antielectrons were common. Neutrinos and antineutrinos were present and they had not yet decoupled from matter so their reactions with nucleons were common. The mass energy of a nucleon is about 938 MeV, so antinucleons were absent (there was not enough energy in the CMB to create them) and nucleons were stable (there was not enough energy to destroy them). They could be thought of as raisins floating in a hot cosmic soup of electrons and neutrinos, their antiparticles, and photons. All of the reactions discussed above were constantly occurring since the average electron and neutrino energy was far greater than the neutron–proton mass difference of about 1 MeV. This served to keep the numbers of neutrons and protons in balance because protons were continually being changed into neutrons, and vice versa.

9.1.2 The second frame ($t = 10^{-1}$ s)

By this time, the density of the Universe had fallen to 3×10^7 times that of water (still mainly in the form of radiation) and its temperature was about 3×10^{10} K. Antielectrons were still relatively common since the average energy of a CMB photon was 3 MeV. Also, the neutrinos had not yet decoupled. However, the electron and neutrino reactions now favor protons over neutrons since protons have lower mass and it is therefore easier to convert a neutron to a proton rather than vice versa. As a result, the nucleon balance was now 62% protons to 38% neutrons.

9.1.3 The third frame ($t = 1$ s)

The density of the Universe was now only 4×10^5 times that of water, so low that neutrinos decoupled from matter to form the cosmic neutrino background discussed earlier. The two *inverse* reactions are therefore ineffective. Also, the temperature has fallen to 10^{10} K corresponding to an average photon energy in the CMB of 1 MeV, barely enough to produce antielectrons. The number of them falls as electron–positron annihilation begins to take over. The p–n balance has now fallen to 76% protons and 24% neutrons but it is still too hot for nuclei to form.

9.1.4 The fourth frame ($t = 10$ s)

The temperature of the Universe in this frame is only 3×10^9 K, corresponding to an average photon energy of roughly 0.3 MeV. This is too small to create pairs so the antielectrons disappear very rapidly. The p–n balance has fallen to 83% protons and 17% neutrons. Nuclei like helium-4 would be stable if they could be produced. However, formation of He is delayed due to the fragility of deuterium (H-2) which must first be formed. This is known as the *deuterium bottleneck*.

9.1.5 The fifth frame ($t = 100$ s)

The temperature has fallen to 10^9 K and the p–n ratio is now 87% protons and 13% neutrons. The deuterium bottleneck has been passed and the elements up to Be are rapidly cooked up. The fraction of He-4 by weight is twice the neutron percentage, or 26%, in agreement with present-day measurements of He abundance. The decay of any remaining free neutrons begins to become important. (The half-life of the free neutron is about 10 min, which means that 50% of a sample of neutrons will decay to protons in a time of 10 min.) As a result, about 10% decay every 100 s. Elements heavier than Be are not made due to the stability gaps at mass 5 and 8 discussed above and the continued rapid cooling of the Universe (see figure 9.1). Note also that the Be isotope produced is radioactive and decays to Li.

9.1.6 Later frames

The time of nucleosynthesis was now over and the hectic pace of the evolution of the Universe slowed down. As mentioned earlier, the Universe became matter dominated at $t = 50\,000$ years and photons decoupled at $t = 400\,000$ years. The *Dark Ages* of cosmology follow that event. During this period, stars had not yet formed and the

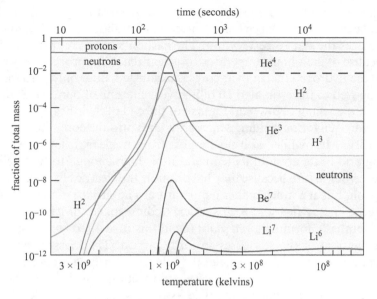

Figure 9.1. The nucleosynthesis era.

only radiation visible would have been that emitted from heated neutral hydrogen gas. Eventually, however, galaxies and stars formed via the gravitational collapse of the density fluctuations in the CMB previously mentioned. (The origin of these fluctuations will be discussed later on.) This is estimated to occur at about 200–400 million ($2-4 \times 10^8$) years after the Big Bang, initiating the era know as *reionization* which extends to the present. Galaxies born at only about 4×10^8 years after the Big Bang have already been seen using the HST, but that is at the limit of its sensitivity. The James Webb Telescope (launch date currently estimated as 2018) will have the power to probe this era in detail and observe the formation of the earliest galaxies.

9.2 The birth and death of stars

Stars form from gravitational collapse of local high-density regions. If the Universe were perfectly uniform, the gravitational attraction on any particle of matter would be the same in all directions and nothing would happen. Fortunately, that is not the case. Local high-density regions attract more matter to them and therefore grow in mass while shrinking in size and heating up due to compression. Eventually, the centers of these giant gas clouds become hot and dense enough to initiate nuclear fusion. Fusion may occur when two atomic nuclei come close enough together to react with one another via the *nuclear force* (more about that later), but atomic nuclei are electrically positively charged and therefore repel each other. At a high enough temperature, however, their kinetic energy is large enough to overcome this repulsion and fusion can occur with the release of energy. The outgoing fusion energy balances the collapse of the gas cloud and eventually a stable star is formed. In the early high-density Universe the initial gas clouds are predicted to have been very large, as were the stars that formed from them. The temperature of these massive stars is very high: large amounts of energy streaming out from the core of the star are necessary to balance the gravitational attraction of all that mass and that in turn requires the star to be very hot. The earliest stars were therefore *blue super-giants*. Because of their high temperature, nuclear reactions proceed very rapidly and quite soon the hydrogen fuel in their cores is exhausted, after only perhaps 10 million years as opposed to the estimated 10 billion year lifetime of our own Sun. Blue stars 'live fast and die young'. However, that is not the end of the story by any means. The fusion of hydrogen forms helium, which gravitates into the core of the star due to its higher density. Meanwhile, with no support from nuclear fusion, the star resumes contracting and heats up until its temperature is high enough to fuse helium nuclei (which are harder to fuse because they have double the electric charge). This re-ignites nuclear fusion, but at a much higher temperature. The surface of the star is blown off by the intense energy flux and forms an expanding sphere of gas which cools as it expands, eventually forming a *red giant* (or in this case a red super-giant) star. The helium in the core of the star fuses to form carbon. The mass-8 gap that was so effective in preventing elements heavier than Be from forming is bypassed in the core of a star by the *triple-alpha* process, a very rare reaction in which three helium nuclei fuse together nearly simultaneously. The astrophysicist Fred Hoyle suggested this process and noted that it requires the existence of a very special state in carbon-12 (the *Hoyle state*) which was soon discovered and found to have exactly the required properties. Once carbon-12 is formed, fusion processes can produce other elements near carbon. Yet again, however, the fuel in the core of the star is eventually exhausted and it collapses and heats up until carbon can be fused. This happens much faster than in the case of hydrogen fusing to helium, since the temperatures are so much higher. In the case of our Sun, for example, the *helium fusing* stage is estimated to last only about 100 000 years compared to 10 billion years for hydrogen fusion. The Sun's mass is too low to allow further fusion reactions to be ignited. However, in a blue super-giant star this sequence continues as the core of the star fuses carbon, then oxygen, neon, magnesium, silicon, and finally sulfur to form an iron core. Each successive stage is faster than the previous one. Calculations indicate that the remnants of these various fusion stages surround the core in a shell structure like that of an onion. The sequence stops at iron because the nuclear physics is such that fusion of iron does not produce energy. Instead, energy would be absorbed to fuse iron so it acts, in a way, like a fusion refrigerator. See figure 9.2.

The star now has no way to prevent its further collapse and heating. Gravity wins and it rapidly implodes and then bounces back into a *supernova explosion* distributing all the 'ashes' from its several fusing stages into the local environment. The energy released in this process is enormous, such that a supernova can for a while outshine all the billions of stars in its galaxy. Nuclear astrophysicists have been quite successful at modeling this process and showing that it results in the observed abundances of elements up to zinc (although research on this topic is continuing), see figure 9.3. They have also shown that elements heavier than zinc are produced in the violent supernova explosion itself, although some details remain to be understood. The remarkable conclusion of this research is that all the atoms in your body, apart from hydrogen, were once at the center of a supernova. You are quite literally stardust!

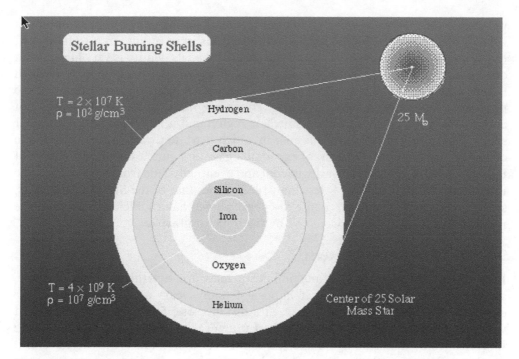

Figure 9.2. Calculation of the structure of a 25 solar mass star at the end of nuclear 'burning', before its collapse and destruction in a supernova explosion.

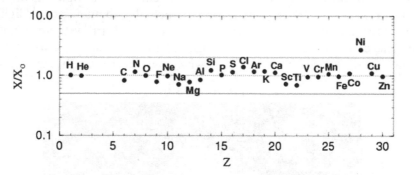

Figure 9.3. A modern calculation of the abundance of elements up through zinc (Zn), as a function of the number of protons in their nuclei (Z). The ratio of the calculated to the experimental value is shown. The lightest elements shown (H, He) were produced in the Big Bang.

Another prediction is that the very oldest stars were made mostly of hydrogen and helium, with little or no contamination by heavier elements. This is not the case for the Sun, since the Fraunhofer lines show the presence of heavy elements, but it is true for the oldest stars that have been observed. Our Sun, then, was formed after the cosmic environment was polluted by heavy elements, which is not surprising since the age of the Solar System is only about 5 billion years compared with 13.8 billion years for the Universe. Finally, it should be mentioned that our Sun will not end its life in a supernova explosion. It is not massive enough to proceed beyond the red-giant stage. Instead, when its helium fuel runs out its core will collapse into a *white dwarf* supported by the repulsion of electrons in its atoms. In a white dwarf, the atoms are close enough to touch each other and so its density is very high, about one million times that of water. It is also initially very hot due to compression, which is why it appears to be blue-white in color. The mainly hydrogen atmosphere of the progenitor star is ejected by the energy streaming from its core, but there is not enough mass to confine it to the immediate region and it blows off into interstellar space. The result is a *planetary nebula* in which a white dwarf can be seen at the center of a ring (actually a shell) of expanding gas, see figure 9.4. Over the course of billions of years, the white dwarf will cool and eventually fade from view. This is the fate of any star whose mass is less than 10–20 times that of the Sun. Heavier stars, such as the early blue super-giants discussed above, have sufficient mass that the repulsion of their atoms is overcome by the force of gravity.

9.3 The size of the Universe

Because the speed of light is not infinite a telescope acts like a time machine, allowing us to probe events that happened many years ago. For example, we currently see the Andromeda Galaxy, at a distance of 2.5 million light years, as it was about 2.5 million years ago. The best present determination of the age of the Universe (to be discussed again later) is about 13.8 billion (1.38×10^{10}) years, so it would at first seem that nothing should be seen beyond a distance greater than 1.38×10^{10} light years. However, this simple calculation does not take into account the expansion of the Universe after light has been emitted and while it is travelling to Earth. The radius of the *observable Universe* is actually about 4.7×10^{10} light years. Light from further distances could not have reached the Earth yet and so nothing in the Universe can be observed at further distances. In fact, nothing currently beyond that distance can ever be observed. To see why, consider the fact that the speed of the

Figure 9.4. The famous 'Helix' or 'Eye of God' planetary nebula (NGC7293). The white dwarf at its center can be seen. The other stars are background stars, not associated with the nebula. Image courtesy NASA, NOAO, ESA and The Hubble Helix Team.

expansion of the Universe increases with distance (the Hubble law). At some distance (the edge of the visible Universe), this recession speed exceeds the speed of light and so light from further distances can never reach us. Note, however, that this is true for observers on any galaxy, including those at the limits of our observable Universe who can see beyond what is visible to us. As a result, we do not really know the actual size of the Universe, which may in fact be infinite! (It is sometimes objected that the special theory of relativity implies that nothing can go faster than the speed of light. However, this applies only to objects moving through space, not to the expansion of space itself.)

Chapter 10

The very early Big Bang

10.1 The four forces of nature

Most of us are familiar to some extent with three forces: gravity, electricity, and magnetism. However, James Clerk Maxwell, in the late 19th century, showed that the latter two forces are actually just two manifestations of a single force we now call electromagnetism. This was the situation until studies of the physics of atomic nuclei in the early 20th century revealed the existence of two other forces, the strong and weak nuclear forces, the effects of which are only apparent on the subatomic level. In the following, we discuss the properties of these four forces of nature.

- **Gravity:** Contrary to common experience, gravity is actually the weakest force in nature by far. On a scale where the strength of electromagnetism is one, the strength of gravity is only about 10^{-36}! Our perception that gravity is strong results from two of its properties. First of all, gravity is always attractive so that the effects of the many, many atoms in a typical object add together. Second, gravity is a *long range* force. It follows an inverse square law and can act over very large distances such as those between neighboring galaxies.
- **Electromagnetism:** This force is also long range and, as mentioned, it is incredibly stronger than gravity. However, electrical charges come in two *flavors* (+/−) and most atoms in the Universe are electrically *neutral*, i.e. they have equal number of positive and negative charges. This means that the effects of electromagnetism are only apparent in cases where the charges are unbalanced, as for example in a lightning storm.
- **The strong nuclear force:** Studies of the atomic nucleus in the 1930s showed that there must be a force other than the very weak force of gravity to hold it together. This force clearly must be stronger than electromagnetism since all the protons in the nucleus repel each other and would otherwise completely disrupt it. In fact its strength is about 100 times that of electromagnetism.

The explanation of the fact that we do not directly perceive this very strong force is that it has a very short range, not much more than the radius of a proton or neutron.

- **The weak nuclear force:** This force is responsible for the radioactive decay of atomic nuclei. Its strength is about 10^{-3} that of electromagnetism so, despite its name, it is many orders of magnitude stronger than gravity. However, its range is much less than even the radius of a proton.

10.2 The quantum nature of forces

In his study of the nature of electromagnetism Richard Feynman showed that, on the deepest level, this force results from the exchange of photons between charged objects. He built on the *uncertainty principle*, introduced by Werner Heisenberg in 1927, which is the foundation of quantum mechanics. In its energy–time formulation, this principle can be expressed in the form of a simple equation:

$$\Delta E \cdot \Delta t = h/4\pi.$$

Here h is Planck's constant and the two quantities on the left-hand side of the equation are as follows. ΔE is an *uncertainty* in the energy of a system and Δt is a time interval. One way to look at this is that there is a fundamental limit on the precision to which the energy of a system can be measured and that the longer the time interval of the measurement, the smaller the uncertainty in energy will be since the product of these two quantities is a constant. (The equation above is actually written in the so-called *quantum limit*. Since it is always possible to make a poor measurement, the equality sign is generally replaced by 'greater than or equal to'.) Because h is very small, this fundamental limit is usually (but not always) far smaller than any realizable experimental precision, but the principle itself has been verified in many different situations. However, there is another way to look at this equation in the light of the *law of conservation of energy*, which states that energy may be changed from one form to another but the total amount of energy always remains constant. This is usually thought of as a fundamental law of nature, but the uncertainty principle provides a way around it. Specifically, the energy of a system may briefly increase by an amount ΔE so long as this extra energy disappears in a time Δt given by the above equation. One analogy is with a bank having a very unusual lending rule: the more money you borrow, the sooner you have to pay it back. In the context of the uncertainty principle, you can 'borrow' any amount of energy from the Universe as long as you pay it back in the appropriate amount of time. Returning now to Feynman's picture of electromagnetism, the force between two charged objects is transmitted by the exchange of photons between them. This is often illustrated by analogy to two skaters moving parallel to each other on a sheet of ice. One carries a heavy ball, which she throws to her partner. The momentum of the ball causes him to deviate from his path after he catches it. Similarly, Newton's third law says that she will also recoil after releasing it. The net result is that the skaters appear to repel each other. Of course, electromagnetism generates attractive as well as repulsive forces so the analogy breaks down in detail. Nevertheless, it gives some feeling for the concepts involved. Note, however, that in Feynman's case the photon does not exist before it is exchanged. Since a photon always has some energy its creation involves a violation of the law of conservation of energy, so it must disappear in a time given by the above equation. Since every photon moves at the speed of light, the distance it can travel before it is absorbed by another charged object is directly determined by the amount of energy it contains—the more energetic the photon, the smaller its range. The net effect is that the force becomes weaker with distance, but its range can be arbitrarily large as the borrowed energy approaches zero. It turns out that this exactly generates an inverse-square-law force. Feynman called his theory *quantum electrodynamics* (QED), and it is among the most accurate theories in physics that have ever been constructed. Some of its predictions have been verified at the level of only a few parts per trillion!

It is now believed that all four forces are transmitted by the exchange of a *force quantum*. In the case of gravity, this particle is called the *graviton*. It is presumed to have zero rest mass just like the photon since that is required to generate an inverse-square-law force as discussed further below. However, due to the extreme weakness of gravity, the graviton has not yet been observed and remains a hypothetical particle. The quantum of the strong force is the *gluon*. Free gluons have also not been directly observed, but for the exact opposite of the case of the graviton. The strong force is so strong that free gluons cannot exist, as will also be discussed further below. In the present state of the Universe, they must remain *bound* inside other particles. Gluons have another property that is crucial for the understanding of the force they generate. Unlike the photon and the graviton, they have a rest mass that is approximately 0.2 times the mass of a proton. What this means is that there is a minimum amount of energy (its rest-mass energy) that must be borrowed to create a gluon in the first place. This puts an upper limit on the range of the force, which can be calculated by assuming that the gluon travels at most at the speed of light. This distance turns out to be about 1×10^{-15} m, in agreement with the observed very short range of the strong force. Finally, the weak nuclear force is transmitted by three force-carrying particles discovered in 1983. These are the $W^\pm$ and the Z^0, where the superscripts indicate their electrical charge. The mass of each of these particles is about 100 times the proton mass and the associated range ($\sim 2 \times 10^{-18}$ m) is far smaller than the size of a proton.

10.3 The unification of forces

Ever since Maxwell showed that electricity and magnetism were two aspects of the same underlying electromagnetic force, theoretical physicists have asked whether all the forces in nature could be understood in the context of a single comprehensive theory. After he formulated the general theory of relativity, Einstein spent much of the rest of his life trying to produce such a *unified field theory*, with little success. One of the more obvious problems is how to account for the very different strengths of the four forces. In the 1970s, however, Sheldon Glashow, Abdus Salam, and Steven Weinberg succeeded in unifying electromagnetism and the weak nuclear force into an *electroweak* interaction. They were awarded the 1979 Nobel Prize in physics for this remarkable accomplishment. One key prediction of the theory was the existence of the $W^{\pm}$ and Z^0 *force-carrying particles*, subsequently discovered in 1983. These play the same role for the weak force that the photon plays for the electromagnetic force. In the electroweak theory, the two forces appear to be very different at everyday low energies but merge into a single interaction at a *unification energy* equal to about 100 GeV, the rest-mass energy of the Z^0. The equations describing this are complicated, but the basic idea is that the uncertainty principle allows for the production of *virtual* Z^0 particles by 'borrowing' energy from the Universe. They are called virtual since they exist only for the brief amount of time allowed by the uncertainty principle. This borrowing becomes easier at higher energies since less energy must be borrowed and at the unification energy Z^0s become common since no extra energy is required to produce them. Under these conditions, quantum mechanics predicts that Z^0s and photons 'mix' together to form a combined particle that transmits the electroweak interaction. Also, electrons and neutrinos mix together to form combined particles that are neither electrons nor neutrinos. These concepts have been verified in great detail by experiments at particle accelerators, such as the Tevatron at Fermilab, which could reach sufficiently high energies. The relevance for cosmology is that at a very early stage of the Big Bang, when the temperature of the Universe ($\sim 10^{15}$ K) corresponds to an energy above the unification energy of 100 GeV, the electromagnetic and weak forces are combined into one single, unified interaction. At about 10^{-12} s after the Big Bang the temperature drops below this critical value and the separate forces begin to emerge.

Electroweak theory provides a framework for the discussion of progress toward unification of the electroweak and strong nuclear forces, which will be deferred until the next chapter. The unification with gravity, however, presents its own problems. Einstein's general theory of relativity describes gravity as resulting from the local deformation of the *space-time continuum* by a mass. In order to describe it in the same way as the electroweak interaction, a *quantum theory of gravity* is necessary, but no such theory currently exists. One of the main reasons (apart from the extreme weakness of gravity) is that a quantum theory involves the exchange of a force-carrying particle (the graviton). At high energies this is not compatible with a smooth space-time continuum, which is instead roiled by fluctuations that become increasingly violent as the energy increases. The result has been dubbed the *quantum foam* (see figure 10.1).

Attempts to deal with this situation have not to date been successful, although some progress has been made as will be discussed further in a later chapter. However, on very general grounds it can be argued that unification with gravity will occur at about the *Planck time*, equal to about 10^{-43} s, which is derived from a particular combination of Newton's gravitational constant and Planck's constant (which is characteristic of quantum theories). The temperature of the Universe at the Planck time after the Big Bang was approximately 10^{32} K. At this temperature it is presumed that all four forces were unified, but as the Universe cooled gravity split off from the other forces.

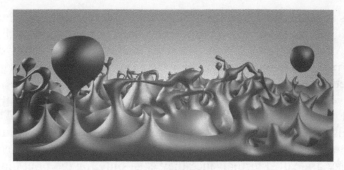

Figure 10.1. Illustration of the quantum foam. The continual creation and annihilation of virtual particles produces violent fluctuations that disrupt the smooth space-time continuum required by Einstein's general theory of relativity. X-ray: NASA/CXC/FIT/E. Perlman; Illustration: CXC/M. Weiss.

10.4 The quark model

From the time of the discovery of the atomic nucleus by Ernest Rutherford in 1911 until the development of powerful particle accelerators in the 1950s, it appeared that all the matter in the Universe could be completely described using only a small number of *elementary particles*, beginning with the proton and the electron which are the fundamental constituents of all atoms. However, measurements of the masses of atoms in the 1920s resulted in the discovery that atoms of a given element may have different masses (as mentioned above, these are called the *isotopes* of the element). The measured masses were always approximately an integer multiple of the proton mass. In 1932, Chadwick proposed that the solution to this conundrum involved the existence of yet another elementary particle, the *neutron*, which is very similar to a proton except that it has zero electrical charge. He subsequently performed a series of measurements demonstrating the validity of this hypothesis. In the same year, Carl Anderson discovered the positron, the antiparticle of the electron, which had been predicted in 1928 by Paul Dirac. The Dirac theory also predicted that every particle had its own corresponding antiparticle (although the antiproton and antineutron were not actually discovered until much later). In 1930, Wolfgang Pauli proposed the existence of yet another particle, the neutrino, in order to explain features of the radioactive decay of certain elements. The neutrino was eventually discovered by Reines and Cowan and their collaborators in 1956. Reines and Cowen were awarded the Nobel Prize in physics almost forty years later, in 1995, for this work. The number of particles was growing, but not so much that the idea of elementary particles was questioned. Then, with the development of high-energy accelerators in the 1950s, the situation changed dramatically. Experiments using these accelerators eventually resulted in the discovery of hundreds of different types of particles, each with their own peculiar characteristics. The field of elementary particle physics came to resemble taxonomy in biology: the discovery and classification of new species. It was difficult to see how anything in this zoo of particles could be considered to be elementary and in fact the name of the field was changed to 'high-energy physics' to reflect this. Something was clearly missing: an underlying principle to explain all these objects. Then, in 1962, Murray Gell-Mann introduced the *eightfold way*, a classification scheme that ordered the particles known at the time. This led him to predict the existence of a yet-undiscovered particle, now called the Ω^-, which was subsequently found and shown to have the predicted properties. Gell-Mann was awarded the 1969 Nobel Prize in physics for this work, which became the basis of an even more comprehensive theory known as the *quark model*, first proposed by Gell-Mann and George Zweig in 1964. In this model, the known heavy particles or *hadrons* (a classification which excludes light particles such as electrons, neutrinos, and photons) are composed of truly fundamental particles which Gell-Mann called *quarks* after a phrase in *Finnegan's Wake* by James Joyce. These objects had some very unusual properties, such as the fact that they were supposed to have electrical charges that are integer multiples of 1/3 the charge of the electron. Fractionally charged particles had never been observed, so at first the quark model created an industry of high-energy physicists looking for such bizarre objects. Nothing of the sort was ever found and they have never been observed even up to the present day despite the fact that the quark model is the accepted theory of hadrons. (The reason for this will be discussed further below.) Another feature of the theory was that it required the existence of three different types of quarks. Two of these were called the *up* (u) and *down* (d) quarks, having charges of +2/3 and −1/3, respectively. The proton and neutron are made of specific combinations of three of these quarks: uud and udd, respectively. In addition, antiquarks having the opposite charge exist. For example, the *quark structure* of the antiproton is ūūd̄, where the bar above a particle symbol indicates that it is an antiparticle. The third type of quark, the *strange* quark (s) having electrical charge −1/3, was necessary to explain the decay properties of certain elementary particles which took much longer to decay to lighter particles than expected. These were called *strange particles* and their quark structure contains strange quarks. For example, the quark structure of the Ω^- is sss. The various types of quarks are called the quark *flavors*. The hadrons have another property that allows them to be classified into two distinct groups and that is their *spin*. Spin is a property that is related to rotation, as might be guessed from the name, but it is much more abstract. It turns out that particles may have either integer or half-integer multiples of the fundamental unit of spin $h/2\pi$, the same quantity that appears in Heisenberg's uncertainty principle. These two classes are called *bosons* (integer spin) and *fermions* (half-integer spin), named after Satyendra Bose and Enrico Fermi who discussed their properties. Quarks are fermions, so particles like the proton, neutron, Ω^-, etc, which are formed from three quarks, must have half-integer spin according to the rules of quantum mechanics. The proton and neutron, for example, have spin $\frac{1}{2}$ while the Ω^- has spin 3/2. Note that the light particles (*leptons*) also have spin, as do *force-carrying* particles such as the photon. With regard to the hadrons, however, there exists a class of particle called *mesons*, for historical reasons having to do with the fact that the first-known members had masses midway between those of the electron and the proton. These particles are all bosons and their quark structure contains a combination of a quark and an antiquark. Quarks are baryons (see above) and mesons therefore have baryon-number zero since they are combinations of a baryon with an antibaryon.

The other hadrons, made up of three quarks, are all baryons (the baryon number of a quark is 1/3, so three of them yield a baryon number of one). (Note that very recent work at the LHC has given some evidence for *pentaquark* objects consisting of five quarks, but the jury is still out on whether such exotic particles actually exist.)

At first the quark model was regarded as a pretty mathematical construct but the reality of the underlying quarks was doubted, especially since nobody had ever seen a free quark. That all changed dramatically in 1974 when a particle that did not fit into the eightfold way was discovered, nearly simultaneously, by groups at Brookhaven National Laboratory (BNL) and the Stanford Linear Accelerator Center (SLAC). Known as the J/ψ particle, it was soon realized that it signaled the existence of a fourth quark flavor, now called *charm*. Actually, this quark had been predicted in 1970 by Sheldon Glashow on the basis of some anomalies in the behavior of the known particles, but his suggestion was ignored until the J/ψ appeared. The electrical charge of the charm quark is +2/3 and together with the strange quark it forms a second *generation* of quarks parallel to the first generation (u, d). The reason that the charm quark had not been discovered earlier is the fact that it has significantly more mass than the other quarks and so it takes higher energy accelerators to produce particles containing charm. At this point, the search was on for other 'charmed' particles. An extension of the quark model predicted the properties of all these particles and they were all eventually found. Not only that, but a third, even more massive generation of quarks, the *bottom* quark (b, charge = −1/3) and the *top* quark (t, charge = +2/3) were discovered, and essentially all the particles containing these quarks have been seen. For a time it seemed that the number of *elementary* quarks might be growing without bounds, similar to what happened to the number of particles in the 1960s. However, there are now good theoretical and experimental grounds to believe that there are only three generations of quarks.

10.5 The leptons

In parallel to what was happening with the heavy particles, the family of leptons was also growing. The first addition was the *muon* (symbol: μ^-) discovered by Carl Anderson and Seth Neddermeyer in 1936. It appeared to be identical to the electron, having the same electrical charge and spin, but it was 210 times heavier. Another new lepton, the *tau* particle (symbol: τ^-), was discovered in 1974. Found to be 3500 times (!) as massive as the electron, it also has the same charge and spin as the electron. The τ is almost twice as heavy as a proton, so why is it classified as a lepton? It turns out that the real distinguishing feature of leptons is that they, unlike hadrons, do not 'feel' the strong nuclear force and only interact via the electroweak interaction. The number of neutrinos was also proliferating and it was discovered that each electron-like particle has its own unique neutrino, established by looking at reactions involving them. As a result, neutrinos are given a subscript indicating which particle they are associated with (ν_e, ν_μ, ν_τ) and the leptons are grouped into three generations just like the quarks. Until recently it was assumed that all the neutrinos had zero mass, but this is now known to be wrong. They have a very small mass, but their actual masses have not yet been determined. Finally, as with the quarks, all the leptons have their corresponding antileptons.

10.6 The gluons

As previously discussed, gluons are the force-transmitting particles of the strong nuclear force, which holds protons and neutrons in the nucleus and also binds the quarks together to produce hadrons. Theoretical investigations of this force ultimately revealed the reason why free quarks (and gluons) have never been seen. It turns out that, unlike the other forces we have discussed, the strength of the strong force actually *increases* with distance. The gluon interaction between two quarks is more like the force exerted by a rubber band—the more it stretches the greater the force it generates. Of course, a rubber band will ultimately break if it is stretched too far. That also happens with gluons, but instead of the quarks coming free the energy of the interaction is large enough to create a particle–antiparticle pair. This unique property of the strong nuclear force is modeled by associating a *color charge* (either blue, green, or red) with quarks and gluons. The analogy here is to the result of mixing primary colors to make white, but of course the color in this case is just an abstract property of the particles. As an example of the principle, the three quarks in a baryon each have one of the three primary colors associated with them, so the baryon is white (or colorless). Gluon interactions change the colors of the individual quarks, but always in such a way as to preserve this property since the main requirement of the theory is that *observable particles in the present-day Universe must all be colorless*. In the case of a meson, which contains a quark and an antiquark, the associated color and anticolor are such as to produce a colorless particle. Gluons are also colored, which means that they feel the strong nuclear force. This is different from the case of the photon, which is electrically neutral and therefore does not feel the electromagnetic force it generates. This property of the gluons accounts for the very unusual nature of the gluonic force. Again, however, free colored objects cannot exist in the present Universe and free gluons are never observed. All of this sounds very abstract, and it is, but it has been incorporated into a theory that is similar to Feynman's QED. Since the theory involves color charges, it has been called *quantum chromodynamics* (QCD). Though many of the predictions of QCD have been verified, the theory itself is quite complicated and it can be formulated in several different ways. As a result, it is not yet as well established as QED and more work is necessary before it will be possible to completely understand all its implications.

10.7 The standard model of high-energy physics

All of the phenomena discussed above, the three generations of quarks and leptons as well as the force-carrying particles and their properties, together make up the *standard model of high-energy physics*, illustrated in figure 10.2. Actually, there was until recently one remaining unobserved particle predicted by the theory: the *Higgs boson* which was necessary to account for the fact that particles have mass (as will be discussed further below). This last missing object was discovered in 2012.

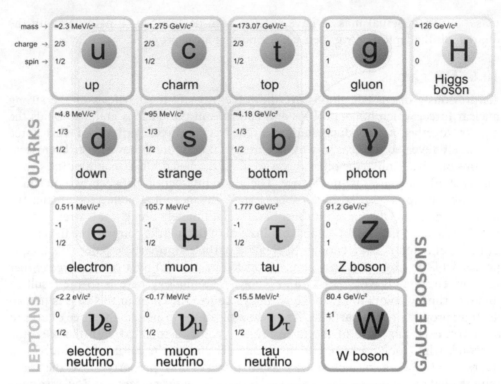

Figure 10.2. The standard model of high-energy physics. This Standard Model of Elementary Particles image has been obtained by the authors from the Wikimedia website https://en.wikipedia.org/wiki/Standard_Model where the author is stated to be MissMJ and it was made available under a CC BY 3.0 licence.

10.8 The history of the Universe: the early frames

With this background in the standard model, it is possible to extend the description of the history and evolution of the Universe given before to much earlier times after the Big Bang:

- **The Planck era** ($t = 10^{-43}$ s): Until this time all of the forces of the Universe, including gravity, were presumably unified into one *superforce* and space-time itself was a *quantum foam*. A viable quantum theory of gravity is necessary to describe the condition of the Universe at this and earlier times, but none yet exists and so the descriptions are speculative and to some extent metaphysical. However, as the Universe cooled, gravity separated from the remaining forces and the quantum foam evolved into a space-time continuum.

- **Grand unification** ($t = 10^{-34}$ s; $T = 10^{27}$ K): The unification of the strong and electroweak forces is called *grand unification*, which is somewhat misleading since gravity remains separate in these theories. At present, there are several grand unified theories that make somewhat different predictions. However, in all of these the quarks, electrons, and neutrinos are mixed together, as are the photons, Z^0, $W^{\pm}$, and gluons. As the temperature of the Universe cooled below 10^{27} K, however, quarks and gluons emerged as the strong force separated from the electroweak force. In addition, a small asymmetry between the numbers of quarks and antiquarks arose at this time, as will be discussed further below. Eventually, the quarks and antiquarks annihilate leaving only a few unpaired survivors that today make up all the matter in the Universe (10^{-9} particles per CMB photon). Another important property of the quarks in this era is that, unlike the case in the present Universe, they do not interact very strongly with each other. As mentioned above, the *color force* between quarks is similar to the force exerted by a rubber band: it grows stronger as the separation between the quarks becomes larger and it becomes very weak at short distances. This property, known as *asymptotic freedom*, implies that quarks and gluons in this early, high-density phase of the Universe form a *quark–gluon plasma* which is *colorful*. In other words, the restriction that objects must be colorless does not yet apply and is imposed only at a much later time when the density of the Universe drops to a value closer to that of a hadron and the quarks and gluons begin to interact very strongly. (Of course, it is well to remember that color charge is an abstract quantity unrelated to what we normally think of as color.)

- **The end of the electroweak era** ($t = 10^{-10}$ s; $T = 10^{15}$ K): At about this time electromagnetism and the weak force, previously united into the electroweak force, part ways. During this transition electrons and neutrinos emerge from their previously mixed phase, as do the photon and Z^0. Light as we know it only exists after this time! In contrast with the earlier eras discussed above, this transition can be studied in the laboratory because the corresponding temperature can be reached (in a small volume of space) with existing high-energy particle accelerators such as the (former) Tevatron at Fermilab. As a result, it has been very well studied and there is little doubt remaining about the state of the Universe at this time.

- **Hadronization of the quark–gluon plasma** ($t = 10^{-6}$ s; $T = 10^{13}$ K): By this time, the density of the Universe has become low enough that the quarks and gluons interact very strongly and (colorless) hadrons begin to be formed out of the quark–gluon plasma. This process is essentially complete at $t = 10^{-4}$ s, after which the Universe is entirely colorless as at present. Also, annihilation of quarks and antiquarks occurs since the temperature is now insufficient to keep their numbers in balance via pair production. As a result only the remaining few unpaired quarks (about one in a billion) are left to form hadrons and the present Universe is made up almost entirely of matter rather than antimatter. The quark–gluon plasma has been studied using the RHIC accelerator at Brookhaven National Laboratory and more recently at the ALICE facility at CERN, which can produce the required temperatures and densities over a (relatively) large region of space by colliding large nuclei such as those of gold or lead that have been accelerated to a sufficiently high energy. One interesting feature that has emerged from these studies is that the plasma seems to behave like a perfect fluid rather than as a gas as was originally expected.

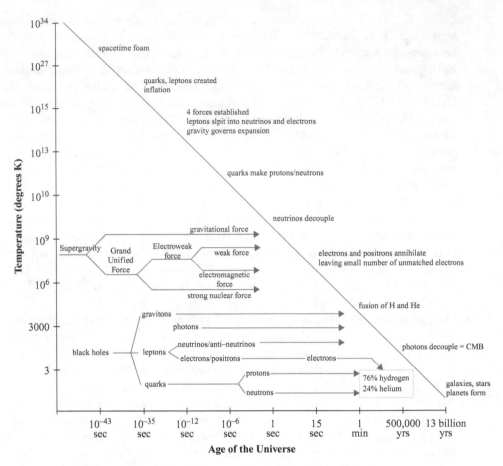

Figure 10.3. A diagram of the evolution of the Universe.

From this time on, the state of the Universe continues to evolve as described in the previous chapter. This entire history is diagrammed in figure 10.3, which summarizes almost everything that modern cosmology has revealed about the evolution of the Universe since the time of the Big Bang. It is truly remarkable that we now have a nearly complete understanding of the state of the Universe at 10^{-10} s and reasonably good information about periods as early as only 10^{-34} s after the Big Bang.

10.9 Why matter rather than antimatter?

As mentioned above, the stuff of the present Universe is predominantly what we call matter. Until recently it was assumed that antimatter was produced only in radioactive decay processes, or in violent astrophysical events resulting in its detection in cosmic rays. However, the PAMELA spacecraft found that a belt of antielectrons (positrons) and antiprotons exists in orbit around the Earth, possibly resulting from the collisions of cosmic rays with the Earth's atmosphere.

Nevertheless, antimatter is extremely rare and it is interesting to discuss why this is the case. In the standard model of high-energy physics, heavy particles (baryons) are assigned a *baryon number*, B, equal to $+1/3$ for a quark and $-1/3$ for an antiquark. Thus, for example, the proton and neutron, which each contain three *constituent* quarks, have $B = 1$. Originally, it appeared that baryon number was a *conserved* quantity. Consider, for example, the process of pair production from photons as previously discussed. In the initial state of this process there are no baryons so $B = 0$. Conservation of baryon number means that B must also equal zero in the final state, which can only happen if a particle ($B = 1$) is produced together with its antiparticle ($B = -1$) and that is what actually happens. However, in 1964 a phenomenon called *CP violation* was observed. A naive interpretation of this process is that antiquarks can very rarely decay into quarks, but not vice versa, so B is not exactly conserved. Certain theories predict that this process will occur with a probability ranging from as much as 10^{-4} (one in ten thousand) to as little as 10^{-12} (one in a trillion) depending on the mechanism and the particles involved, and experiments (such as the BABAR experiment at the former SLAC) were carried out to test them. Soon after the discovery of CP violation, Andrei Sakarov proposed that this was the explanation for the predominance of matter in the Universe. Sakarov, the father of the Soviet atomic bomb, was a prominent Russian nuclear physicist, Soviet dissident, and human rights activist who was awarded the Nobel Peace Prize in 1975. In 1967, he proposed that the asymmetry between matter and antimatter occurred at a very early stage in the history of the Big Bang, during the transition from grand unification at about 10^{-34} s. In order to account for the observed amount of matter, a CP violation probability of about 3×10^{-9} is required and this is within the range of both theory and experiment. Sakarov's point was that pair production in this early phase of the Big Bang would produce equal amounts of matter and antimatter, but CP violation would ultimately result in a small excess of matter. Then the vast majority of the produced quarks annihilate with their corresponding antiquarks, producing photons that ultimately end up in the CMB. The remaining excess quarks find no partner and remain as the matter we see today. This accounts for the 10^{-9} ratio of baryons to photons, as well as the asymmetry between matter and antimatter observed in the present Universe.

Chapter 11

Inflation

11.1 The horizon problem

We have learned that the CMB is very, very uniform. After subtracting the *dipole anisotropy* due to the motion of the Milky Way, the remaining temperature differences from one part of the sky to another are at most a few tens of microkelvins out of an average temperature of 2.7 K. Why is this a problem? Think first about the fact that a telescope is a sort of time machine that allows us to see very distant objects as they were many years ago, due to the fact that the light from these objects has taken a long time to reach us. If we point our telescope in one direction (say towards the east) and look for the most distant objects, we could in principle (although not yet in practice) see them as they were about 13.8 billion years ago at the time of the Big Bang. Turning the telescope to the west, we can observe another set of objects as they were 13.8 billion years ago. Now we come to the nub of the problem. An observer on the west objects cannot see the east objects, since there was not enough time after the Big Bang for light to have travelled such a long way. They are in effect beyond her 'horizon'. Special relativity tells us that no signal can travel through the Universe faster than the speed of light, so no signal could possibly have traveled from the far east to the far west. They are said to be *causally disconnected*. After thinking about this for a while, it should become clear that this was always true from the very beginning. The two opposite sides of our observable Universe have always been causally disconnected. Actually, it is much worse than that. It turns out that any two patches on the sky separated by more than about two times the Moon's diameter have always been causally disconnected from one another. This presents a serious problem for attempts to understand the uniformity of the CMB. If no signal could ever have passed between these causally disconnected regions, how do they 'know' to be at the same temperature to such a remarkable precision? The sky should instead consist of a patchwork quilt of many areas having different temperatures. There is an alternative, of course. Perhaps the initial conditions of the Big Bang were set up such that the temperature of the CMB was the same in all directions. This *fine tuning* hypothesis is not very satisfactory for cosmologists, who would like to understand why this was so—hence the *horizon problem*.

11.2 The flatness problem

The most recent observations of the curvature of the Universe have shown that it is very nearly 'flat'. The curvature parameter $\Omega = 1.02 \pm 0.02$ from WMAP data (see the next chapter). This seems to be a very unlikely result since it requires an exact balance between matter and gravitational energy. In fact, this *flatness problem* is much worse than it initially appears since it turns out that Ω evolves with time. If Ω is even slightly greater (smaller) than one, it becomes progressively greater (smaller) as time goes on. This is known as *unstable equilibrium*, and an analogy is to a pencil exactly balanced on its point: any disturbance will cause the pencil to fall. Calculations show that if the difference $\Omega - 1 = 0.01$ at present, then it must have been only 10^{-25} at 1 ns after the Big Bang and even smaller earlier on. This truly extreme fine tuning of the initial state of the Universe has no explanation within the standard model of the Big Bang.

11.3 The smoothness problem

If the Universe is as flat and uniform as we have seen it to be, then how did large-scale structure such as clusters of galaxies ever emerge? The standard explanation is that the fluctuations observed in the CMB acted as seeds for gravity. Without these seeds, there would be no lower or higher density regions for gravity to work on, and matter would never form 'clumps' such as galaxies, stars, or people. Simulations including these fluctuations have been very successful in predicting the observed large-scale structure of the Universe. The *smoothness problem* asks the question: where did those fluctuations in the CMB come from in an otherwise very flat and very uniform Universe?

11.4 The magnetic monopole problem

One remaining problem with the Big Bang theory regards the existence or non-existence of the hypothetical *magnetic monopole* particle. It is common knowledge that all magnets come with two opposite magnetic poles, referred to as North (N) and South (S). If you break a magnet in two, you are left with two magnets each of which has a N and S pole. However, certain theories of particle physics predict the existence of *magnetic monopoles* which have only one isolated magnetic pole. These theories also tend to be leading candidates for quantum theories of gravity and are not easily dismissed. The problem is that magnetic monopoles have never been seen (although there was one possible observation that has never been reproduced). Also, observations of the magnetic fields of galaxies have placed very low limits on the number of magnetic monopoles that might be in them. On the other hand, the theories that predict their existence also predict that many of them should have been produced early-on in the Big Bang, and that they should still be around even now, so where are they? This is perhaps less of a problem than the others discussed above since there are theories that do not predict magnetic monopoles, but it is something that is not understood in the context of the standard Big Bang model.

11.5 Inflation

A truly elegant solution to all four of these problems emerged in the 1980s from the work of Alan Guth and Andrei Linde. It follows from some of the properties of the theories of the unification of forces discussed in the previous chapter. Consider, for example, the electroweak transition in which the weak force separates from the electromagnetic force. This transition is now very well understood since, as previously mentioned, the energy scale at which it occurs can be directly accessed at high-energy accelerators such as the former Tevatron at Fermilab (now retired). During the transition the photon and the Z^0 particle, previously mixed, emerge as separate entities. These two particles are very similar, except that the photon has no rest mass and the Z^0 is very heavy. It turns out that this mass difference is now understood in the context of the *Higgs mechanism*, introduced in the 1960s by the theoretical physicist Peter Higgs, in which the mass of any particle is determined by its interactions with the *Higgs boson*. This is the final particle in the standard model of high-energy physics and it was only discovered in 2012. Peter Higgs was awarded the 2013 Nobel Prize in physics for predicting it. The same mass-generating mechanism also occurs during the QCD transition when quarks emerge as the strong force separates from the electroweak force. Guth and Linde noticed that the Higgs boson generates a high-energy-density state they called the *false vacuum* which has properties similar to Einstein's cosmological constant.

One other ingredient was necessary in order for their theory to work. The early Universe had to remain trapped in the high-density phase for a time during the cooling process as it went through the QCD transition. This type of trapping is at times observed in other phase transitions. For example, when water condenses on a cold window pane, it may sometimes happen that it cools below the temperature at which it normally would freeze into ice crystals. This process is known as *super-cooling*. If the water is disturbed in any way, it immediately freezes into ice crystals with the release of the so-called *heat of crystallization*. There are even commercial *heat packs* that make use of the supercooling of a chemical solution. A similar phenomenon called *superheating* can at times be observed when distilled water is heated in a microwave oven. The temperature of the water can rise above the

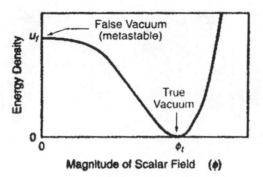

Figure 11.1. Illustration of the false vacuum. The scalar field is generated by the Higgs boson (or a similar particle). The energy density of the Universe is reduced when quarks gain mass. The false vacuum is said to be metastable since it disappears when the Universe transitions to the lower-density true vacuum state.

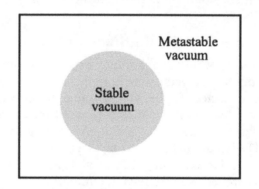

Figure 11.2. Formation of a bubble of true vacuum in the metastable false vacuum.

temperature at which it would normally become steam. This can be dangerous since disturbing the water as it is removed from the oven may then result in very rapid boiling. Returning to the QCD phase transition, which occurred at about 10^{-32} s after the Big Bang, Guth and Linde proposed that the Universe remained trapped in the false vacuum as its temperature fell. Then a small 'bubble' of space-time began to make the transition to the lower-energy-density state, which is the *true vacuum* in our present Universe. See figures 11.1 and 11.2.

Because the false vacuum acts like Einstein's cosmological constant, which is similar to repulsive instead of attractive gravity, this bubble grew very rapidly, doubling in size approximately every 10^{-34} s. As a result it doubled about 100 times during the transition, so a region that was only 10^{-15} times the size of a proton was 'inflated' to about 10 cm in diameter. At the same time the energy of crystallization was released, heating the bubble (which makes up our entire Universe) to something like 10^{27} K and initiating the standard Big Bang. This process has become known as *cosmic inflation*. (It should be mentioned at this point that the identification of the

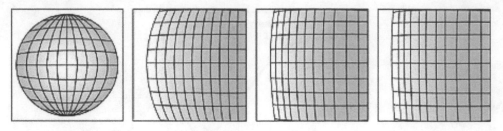

Figure 11.3. The appearance of a sphere after one, two, and three doublings in radius.

recently observed Higgs boson as the source of inflation is somewhat controversial. Guth generally refers to an *inflaton*, which must have similar properties to the Higgs boson but may or may not be identical to it. In addition, there are several different versions of inflation theory that make different predictions.)

11.6 How inflation solves the Big Bang problems

Consider first the horizon problem. Space-time expands so rapidly during inflation that all parts of the bubble Universe were initially in contact with one another and would therefore be expected to be at the same temperature. As mentioned previously, while signals cannot travel through space-time at greater than the speed of light, the much faster expansion of space-time during inflation is allowed by general relativity. The solution to the flatness problem is also straightforward. Whatever curvature the Universe might have had prior to inflation, the massive expansion during this process would stretch it out until it was very nearly flat. Think about blowing up a balloon. After only three or four doublings in size, the curved surface of the balloon has become much flatter, as can be seen in figure 11.3.

After 100 doublings, the curvature of the Universe would correspond to $\Omega - 1 < 10^{-50}$ or so. Therefore, instead of $\Omega = 1$ being an unlikely result, inflation predicts almost exactly this value. The smoothness problem has a more interesting solution. Remember that the Heisenberg uncertainty principle allows for brief violations of the principle of conservation of energy, leading to the fact that particle–antiparticle pairs are always being created and destroyed in the Universe. This production of *virtual pairs* has actually been directly observed via the effect (known as the *Lamb shift*) they have on light emitted from hydrogen atoms. If such a quantum fluctuation occurs during inflation, it will be caught and expanded. Also, the expansion itself will provide enough energy that the fluctuation will not have to disappear to satisfy the uncertainty principle. The result is the density fluctuations observed in the CMB. It is very remarkable to realize, then, that the large-scale structure in our Universe ultimately resulted from tiny quantum fluctuations at the time of inflation. This will be discussed further in the next chapter. Finally, the solution to the magnetic monopole problem is that it is very unlikely that more than a few such objects could have occupied the tiny region of space-time that became our Universe, so the

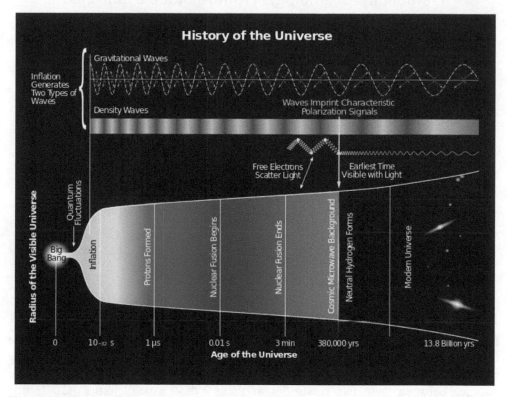

Figure 11.4. The history of the Universe. This History of the Universe image has been obtained by the authors from the Wikipedia website https://en.wikipedia.org/wiki/Chronology_of_the_universe#/media/File:History_ of_the_Universe.svg where it was made available by Yinweichen under a CC BY SA 3.0 licence.

monopole density was vastly diluted by expansion and we should not expect to find any even if very many were initially produced.

The history of the Universe, including inflation, is illustrated in figure 11.4. The role played by *gravitational waves* is discussed in a following chapter.

Chapter 12

Dark energy

12.1 The curvature of space-time

In a previous chapter, the connection between the curvature of space-time and the ultimate fate of the Universe was described. Briefly, in a Universe in which only matter and gravity are important, the curvature of space-time depends on the balance between these two quantities. If there is 'too much' matter ($\Omega > 1$), the curvature will be positive (closed Universe) and the expansion will slow down and ultimately turn into a compression. If there is 'too little' matter ($\Omega > 1$; open Universe) the curvature will be negative and the expansion will continue forever. In the 'Goldilocks' case ($\Omega = 1$) the Universe is flat and the expansion, although never turning into a compression, will eventually stop but only at infinite time. One way to decide among these options is to try to estimate the amount of matter from observations of stars, galaxies, etc. The chapter on dark matter (chapter 8) discusses some of the problems that astronomers have encountered in making the required observations, especially since much of the matter in the Universe is non-baryonic dark matter and not the familiar protons and neutrons that make up the world we see around us. The conclusion from these surveys was that $\Omega \sim 0.3$, including the estimated amount of this dark matter. Taken literally, this would imply that the ultimate fate of the Universe would be perpetual expansion. In that case, matter would eventually thin out as the Universe became more and more extended. Also, the expansion would cool the Universe, which would eventually become too cold to sustain life, resulting in a *Big Freeze*. This is clearly a very important and interesting conclusion if it were correct. However, as mentioned in the previous chapter, the inflation hypothesis predicts the seemingly unlikely Goldilocks Universe. Also, the dark matter estimates are possibly not very accurate, so maybe there is a lot more matter than expected. It is therefore very important to determine the value of Ω in a different way. In the late 1990s, two teams of scientists, one headed by Saul Perlmutter

doi:10.1088/978-1-6817-4100-0ch12

and the other by Brian Schmidt and Adam Riess, developed such a method based on the properties of Type Ia supernovae. These massive and very bright star explosions occur in binary star systems in which one of the stars has already evolved into a white dwarf and the other is expanding into the red-giant phase. Some of the hydrogen gas from the bigger star can then fall onto the surface of the white dwarf and the amount of this *accreted* matter can eventually become so great that it initiates a thermonuclear explosion. The *light curve* (brightness versus time) of the explosion is characteristic and observational studies as well as theoretical calculations have shown that the maximum brightness of all Type Ia supernovae is always the same (apart from some small and well-understood corrections). Furthermore, as mentioned, they are very bright so they can serve as *standard candles* to measure to enormous distances and the speeds of the galaxies containing them can be determined from the red-shifts of the light they emit. Since looking to large distances is equivalent to looking backward in time, the research groups were able to determine the rate of expansion of the Universe as a function of time. Their first results, published in 1998, were incredibly unexpected and surprising. Far from slowing down with time, the rate of expansion of the Universe was actually accelerating! This result has since been repeatedly verified and Perlmutter, Schmidt, and Reiss were awarded the 2011 Nobel Prize in physics for this revolutionary discovery.

12.2 The accelerating universal expansion

The first conclusion from the fact that the expansion rate is accelerating is that there must be something more than just matter and gravity in the Universe. Something must be causing the expansion rate to increase and that something has been called *dark energy*. It is a form of energy, and matter and energy are equivalent according to the special theory of relativity. Therefore, dark energy contributes to the curvature of the Universe and the supernova measurements showed that it actually dominates. In fact, $\Omega_\Lambda = 0.7$, where Λ is the symbol that refers to dark energy. This compares with $\Omega_M = 0.3$, where Ω_M refers to matter. Note that the sum of these two contributions is one, which implies that the Universe is actually flat when dark energy is properly accounted for. As with dark matter, the nature of dark energy is at present not understood. One possibility is that it corresponds to Einstein's cosmological constant. In this case, the rate of acceleration will remain constant and the ultimate fate of the Universe will be as discussed above for an open Universe except that the speed of expansion will be ever increasing so the Universe will thin out even faster. However, another hypothesis is that the amount of dark energy and therefore the rate of acceleration will increase with time. In this scenario, the expansion rate will ultimately become so great that it tears everything, even quarks, apart. This is known as the *Big Rip*. A third possibility is that dark energy may ultimately become smaller, or even reverse sign resulting in a *Big Crunch*. Finally, it should be mentioned that some theorists have suggested that the apparent acceleration of the expansion may be due to a modification of gravity (and hence general relativity) at large distances. However, to date the attempts to incorporate this idea have resulted in theories that are inconsistent with observations and there are yet other suggestions that so far have not proved viable. The correct interpretation will only become clear when dark energy is properly understood.

12.3 Dark energy and the CMB

The supernova observations have shown that $\Omega = 1$, in agreement with the prediction of the inflation theory, but there is a rather large range of uncertainty in this result. It turns out, however, that detailed studies of the fluctuations in the CMB have given very precise values for Ω as well as other important cosmological parameters. In particular, data from the WMAP give $\Omega = 1.02 \pm 0.02$ as mentioned previously. The idea behind this conclusion comes from a study of the locations of the CMB fluctuations. The method is as follows: start with a particular fluctuation and count the number of fluctuations that are $1°$, $2°$, $3°$, etc, away from it. Then, repeat for every fluctuation and add up the entire distribution. The result is a graph that has a big peak around $1°$, plus a few smaller peaks. This is the *acoustic power spectrum* (see figure 12.1) and it results from the effect of gravitational waves in the early Universe moving the fluctuations around (the bang of the Big Bang!). The most important point is that this spectrum is very sensitive to a number of cosmological parameters and provides the best way that we know of to determine them, as illustrated on the WMAP webpage. In addition to the value of Ω, the WMAP data give the age of the Universe as 13.77 ± 0.06 billion years and its current temperature as 2.725248 ± 0.00057 K. Also, the amount of 'normal' matter (atoms, etc) in the Universe (as a fraction of $\Omega = 1$) is 4.6%, in agreement with the prediction of ~4% from nucleosynthesis calculations. Finally, cold dark (non-baryonic) matter makes

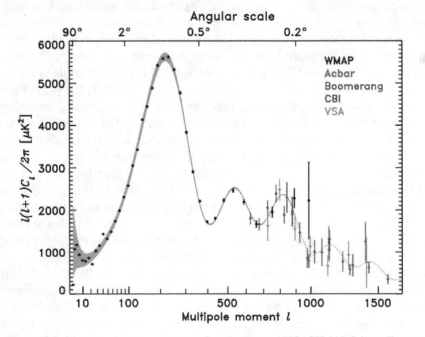

Figure 12.1. The acoustic power spectrum. Image courtesy NASA/WMAP Science Team.

up 24% and dark energy 71.4% of $\Omega = 1$, in agreement with the supernova results and the surveys of matter in the Universe. Measurements using the PLANCK spacecraft, launched by the European Space Agency (ESA), gave even more precise results in 2013. This team gave the age of the Universe as 13.798 ± 0.037 billion years, the amount of 'normal' matter as 4.9%, dark matter as 26.8%, and dark energy as 68.3%.

12.4 Is there a signature of inflation in the CMB?

Theoretical calculations of the Big Bang have shown that gravity waves should have been produced by the tremendously rapid expansion of the Universe during the inflation era. Gravitational radiation is predicted by general relativity, but because of the extreme weakness of the gravitational force it can be produced with detectable intensity only during the most violent astrophysical events. Large instruments are being built to directly detect gravity waves but up to now the search has not been successful. However, the prediction of inflationary theory is that *primordial* gravitational waves will affect the polarization of fluctuations in the CMB in a very specific way, generating so-called *B-mode polarization* that would be a unique signature of inflation. In March 2014, a team using the BICEP2 microwave telescope at the South Pole reported that they had detected the B-mode signal. However, in September of that year the PLANCK team concluded that the signal resulting from light scattering from dust in the Milky Way would be about the same as that measured by BICEP2. Further results from BICEP2 and the BICEP3 telescope currently under construction, as well as new analyses of the existing PLANCK data, are expected to resolve this situation.

Chapter 13

Higher dimensions

13.1 Field theories

Theoretical physicists have found through long experience that the way to address the mathematical description of forces is via *field theory*. A *field* in physics is a collection of numbers, defined at every point in space, which completely describes a force at that point. This has also proven to be the most fruitful method for unifying the various forces found in nature. The first field theory was Maxwell's unification of electricity and magnetism. It turned out that these two seemingly different forces were in fact just two aspects of the same force, completely described by four numbers. These are the *electric potential* and a three-dimensional *vector potential*, which are generally combined into a four-dimensional vector **A** with *components* (A_1, A_2, A_3, A_4). The force of gravity according to Einstein's general theory of relativity is more complicated. It is described by a four-dimensional *metric tensor* (mathematically the *outer product* of two four-dimensional vectors), which has four rows and four columns and looks like this:

$$\begin{vmatrix} g_{11} & g_{12} & g_{13} & g_{14} \\ g_{21} & g_{22} & g_{23} & g_{24} \\ g_{31} & g_{32} & g_{33} & g_{34} \\ g_{41} & g_{42} & g_{43} & g_{44} \end{vmatrix}.$$

It has a total of 16 numbers that have to be defined. However, theory says that corresponding *off-diagonal elements* must be equal: $g_{12} = g_{21}$, $g_{13} = g_{31}$, etc, so in the end only ten numbers are left to be defined. These ten numbers completely describe the curvature of space-time at every point.

13.2 Kaluza–Klein theory

Following Einstein's successful introduction of a four-dimensional metric tensor to describe the curvature of space-time, the German mathematician Theodor Kaluza

doi:10.1088/978-1-6817-4100-0ch13   13-1

decided to investigate the properties of five-dimensional fields, which have 25 numbers to be specified. Again using the property that corresponding off-diagonal elements are identical, this reduces to only 15 independent elements. What Kaluza found, much to his surprise, was that the five-dimensional tensor looks like this:

$$\begin{vmatrix} g_{11} & g_{12} & g_{13} & g_{14} & A_1 \\ g_{21} & g_{22} & g_{23} & g_{24} & A_2 \\ g_{31} & g_{32} & g_{33} & g_{34} & A_3 \\ g_{41} & g_{42} & g_{43} & g_{44} & A_4 \\ A_1 & A_2 & A_3 & A_4 & A_5 \end{vmatrix}.$$

The first four extra row and column elements were actually the same as the corresponding elements of the electromagnetic field. In other words, Kaluza had unified electromagnetism and gravity! In the process, he showed that light could actually be viewed as warpage of a five-dimensional *hyperspace*. The new element A_5 turns out to be a (predicted) *scalar particle*, which, if it exists (i.e. if the corresponding number is non-zero) has the same mathematical properties as Einstein's cosmological constant. We have met an object like this before—the Higg's Boson. It should be understood, however, that Kaluza's theory is *not* a quantum theory of gravity, which as we will see is a much more difficult problem to solve.

13.3 Compactification

Einstein had already identified the fourth dimension as time. Kaluza had to answer the question: where is the fifth dimension? It is obviously invisible to us so it must be very different from the three space dimensions (x, y, z) that we are familiar with. He proposed that it was a *small dimension*, curled up so tightly that it has no visible extent unlike the normal large dimensions (x, y, z, t) that can take on an infinite range of values. One analogy is to a garden hose. Viewed from a distance, the hose looks like a line and appears to have only one dimension: its length. Viewed close up, however, its three-dimensional structure becomes clear: the other two dimensions are curled around in a circle. Think also about an ant moving on the hose. To a viewer at a distance, it appears that the ant can only move forward and backward along the hose, but the ant realizes that it can remain in one place along the length of the hose while travelling around its circumference. Physicists relate this to the concept of *degrees of freedom*. From a distance, the ant appears to have only one degree of freedom: the ability to move back and forth along the hose. The ant, however, actually experiences three degrees of freedom since it can also move around the hose. In his original theory, published in 1921 before Heisenberg introduced his uncertainty principle, Kaluza did not specify the size of the fifth dimension and thought of it as a mathematical point. However, the Swedish physicist Oskar Klein subsequently pointed out in 1926 that the uncertainty principle prevents a small dimension from curling up into a point. Instead, it has to be *compactified* into a region that is the size of the *Planck length*—10^{-35} m. (This is the distance that light travels in the *Planck time* of 10^{-43} s). This modified version, although still not a true quantum theory, is referred to as the Kaluza–Klein theory.

13.4 QED

In order to understand the problems faced by theoretical physicists in attempting to develop a true quantum theory of gravity, it is helpful to start with Richard Feynman's quantum theory of the electromagnetic force which he called QED. As previously mentioned, in this theory the electromagnetic force is transmitted by exchange of the quantum of this force: the photon.

In order to help visualize the process, Feynman invented the so-called *Feynman diagram*. Two examples are shown in figure 13.1.

They represent the exchange of photons (indicated by the curly lines) between two electrons. The first is the 'broken H' diagram corresponding to the exchange of a single photon. The second 'box' diagram represents the more complicated two-photon exchange. These are only two of many, many diagrams that can occur. More complicated examples are shown in figure 13.2.

The first 'tree' diagram represents the production of an electron–positron pair from a (virtual) photon emitted by an electron. In these plots, the horizontal direction represents time and an antiparticle is represented as a particle going backward in time. The second of these is a so-called 'penguin diagram' containing a photon 'loop' connecting between the upper electron before and after the inter-action. In order to make a calculation within QED, it is necessary to evaluate all possible diagrams and sum their effects. However, a serious problem emerged when Feynman tried to do this. It turned out that some diagrams, particularly some loop and penguin diagrams, gave an infinite contribution to the strength of the force. Clearly this cannot be correct: the electromagnetic force is not infinite. However,

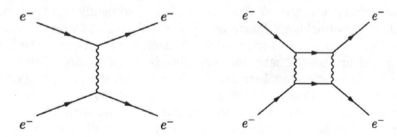

Figure 13.1. Two examples of Feynman diagrams.

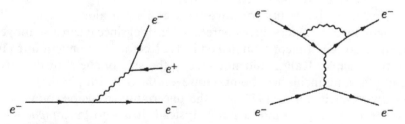

Figure 13.2. Two examples of more complicated Feynman diagrams.

Feynman found that some of the diagrams gave a positive contribution and others gave a negative contribution. When the summation over diagrams was performed in the proper order, the positive and negative infinities cancelled, leaving a finite value. This was called *renormalization* and it is now recognized that all successful quantum field theories have to be renormalizable.

It might seem like this process of cancelling out positive infinities with negative infinities is suspect. Is it possible to accept a theory that does this? It turns out, however, that QED is the most precise theory that has ever been developed. Certain of its predictions have been tested and found to be correct to one part in 100 trillion!

13.5 Quantization of the weak and strong forces

Unification of the weak and electromagnetic forces was accomplished by Sheldon Glashow, Steven Weinberg, and Abdus Salam. They were awarded the 1979 Nobel Prize in Physics for this work. Electroweak theory is a true quantum field theory involving the exchange of $W^{\pm}$ and Z^0 bosons as well as photons. It was eventually shown to be renormalizable by Gerardus t'Hooft and Martinus Veltman, for which they received the 1999 Nobel Prize in physics. It has been very well studied since the energy scale at which it becomes important was directly accessible by existing high-energy accelerators such as the former Tevatron at Fermilab. The strong force has also been quantized in the context of QCD, but there are some major differences compared with the electroweak interaction. The quantum of the strong interaction is the *gluon* but there are actually eight different kinds of gluons which adds substantially to the complexity of the theory. In addition, unlike the photon, which does not interact with itself since it is not electrically charged, the gluons interact very strongly with one another. This leads to the confinement and asymptotic freedom properties of the strong force as discussed in an earlier chapter. Because of this, there exist several different formulations of the theory that apply, for example, at high energy where asymptotic freedom is important or at lower energy where confinement dominates. The energy scale at which the electroweak and strong forces unify is very far beyond what could ever be reached via existing or proposed accelerators, so not all aspects of the theory are as well understood as in the case of electroweak unification. As a result, the state of the Universe at the time of this transition ($\sim 10^{-32}$ s after the Big Bang) is not completely understood and the transition is complicated by the fact that inflation is also occurring at this time.

13.6 Early attempts at a quantum theory of gravity

Returning to the Kaluza–Klein theory, which did successfully unite gravity and electromagnetism, several theoretical physicists attempted to extend it to see if a viable quantum theory of gravity was possible. If successful, this quest would lead to the so-called *theory of everything* (TOE) which would explain all the matter and forces in the Universe. One question then was: how many extra dimensions are needed to describe reality? At the time, there was no guidance on this—up to 26 dimensions and more were investigated. All of these extra dimensions must be compactified into small dimensions to describe our world, which has (for us) only four large dimensions. It turns out that, mathematically, there are many ways to do this. The simplest, however, is to compact the extra dimensions into the hyperspace analog of a sphere—the *Calabi–Yau hypersphere* (see figure 13.3)—and this is usually what is done. However, even this is not very simple since it turns out that there are many ways to compactify the additional dimensions, even if the process is restricted to Calabi–Yau hyperspheres, and each compactification yields a different Universe (see figure 13.4).

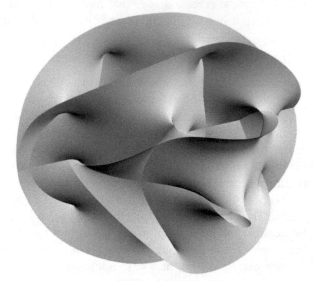

Figure 13.3. An Illustration of a Calabi–Yau hypersphere. This Calabi-Yau image has been obtained by the authors from the Wikipedia website https://simple.wikipedia.org/wiki/Calabi-Yau_manifold where the author is stated to be Lunch and it was made available under a CC BY SA 2.5 licence.

Figure 13.4. Space-time as a matrix of Calabi–Yau hypersperes. Rick Groleau. "Imagining Other Dimensions," NOVA website http://www.pbs.org/wgbh/nova/physics/imagining-other-dimensions.html (accessed August 26, 2015).

Particles and Sparticles

Particle Name	Superpartner Particle	Particle Name	Superpartner Particle
Quark	Squark	Graviton*	Gravitino
Neutrino	Sneutrino	$W^{+/-}$	Wino$^{+/-}$
Electron	Selectron	Z^0	Zino
Muon	Smuon	Photon	Photino
Tau	Stau	Gluon	Gluino
		Higgs* boson	Higgsino

Figure 13.5. Particles and sparticles.

An early leading candidate for a TOE, one of a class of theories know as *supersymmetric theories* (SUSY), was called *supergravity*. Developed in the late 1970s, it used 11 dimensions and appeared to unify the description of forces and matter. This theory predicted the existence of a *super-partner* for every type of known particle (the *sparticles*). The super-partner of a *matter particle* (a quark or lepton) is a *force-carrying particle*, and vice versa (see figure 13.5). These have never been observed, although people are still trying. The recent discovery of the Higg's boson at the LHC has given a new impetus to this search. The mass of the sparticles is determined by that of the Higgs, which was slightly greater than expected. There is hope that the lighter sparticles might be produced during the next run of the LHC at higher energy in mid-2015. But the main difficulty with supergravity was that, although it initially seemed to be renormalizable, in the end some of the infinities just could not be eliminated with the mathematical techniques available at the time.

Chapter 14

String theory

14.1 Particles and 'string'

One of the major obstacles to developing quantum theories of gravity is the fact that general relativity and quantum mechanics do not 'play nicely' with each other. The Universe described by general relativity is a space-time continuum deformed by matter and energy. However, the grand unified force separates from gravity at the Planck time (10^{-43} s after the Big Bang) when the temperature of the Universe is $\sim 10^{32}$ K. At this temperature, quantum mechanics predicts that violent quantum fluctuations play a dominant role. The picture here is that the energy of these fluctuations would be large enough to generate enormous local deformations from a smooth space-time continuum and these deformations would be continually changing as the fluctuations came into being and then disappeared. The result is that the fabric of space-time dissolves into a *quantum foam*. String theory introduces the concept that all the different particle types in nature, of which there are many, result from different vibrations of a single type of *string* that is the size of the Planck length—10^{-35} m. An analogy would be to the different tones emitted by vibrations of a guitar string. Strings can in principle be either *open* (with two free ends) or *closed* (the two ends joined to make a loop). We will return to open strings later on, but if we consider that particles are made from vibrating closed loops of string, then the Feynman diagrams for quantum interactions are replaced by the so-called *pants diagrams*. This is illustrated in figure 14.1, which shows pants diagrams for one- and two-photon exchange between two electrons. Comparing with the corresponding Feynman diagrams shown in the previous chapter, it is clear that pants diagrams cover a larger region of space-time and there is hope that they will average over the violent quantum fluctuations enough to make a consistent quantum theory of gravity.

This idea was first introduced in the 1980s, but the most amazing thing about string theory, and the fact that gave impetus to an incredible amount of research effort in the 1990s, was the discovery that string theory is renormalizable when

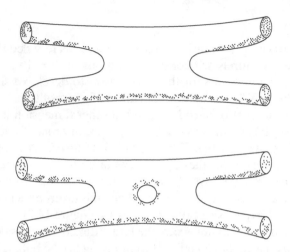

Figure 14.1. 'Pants' diagrams for one- and two-photon exchange.

written down in ten dimensions. All the disastrous infinities in the theory were found to be multiplied by $(N-10)$, where N is the number of dimensions. So far as is known, this is the only theory that specifies an exact number of dimensions for the Universe and the reason for this remarkable behavior is not known. (Actually, 26 dimensions also work, but that version of the theory has some problems that have not yet been worked out, so $N = 10$ *superstring theory* is the current favorite). Given the four large dimensions of space and time, six left-over dimensions of the theory have to be compactified. As mentioned in the previous chapter, there are an enormous number of ways to do this (even if we restrict ourselves to Calabi–Yau hyperspheres) and each yields a different Universe. Sorting through this *string-theory landscape* of 10^{10} to as many as 10^{100} possible Universes to find one that matches ours is beyond the current capabilities of cosmologists. Actually, however, this is not even the biggest problem with string theory, which suffers greatly from the very large energy at which quantum gravity plays an important role. As a point of comparison, the quantum theory of the weak nuclear force was directly probed at the energy of the Tevatron at Fermilab, which is about a factor of ten lower than that of the maximum energy of the current champion, the LHC at CERN. Doing the same for quantum gravity would require an accelerator operating at 100 trillion times that of the LHC, making the predictions of string theory impossible to directly test with any particle accelerator in the near or probably even distant future. Even indirect tests (looking for very small effects at much lower energies) are way beyond current technology, so string theory appears to be untestable. As a result, many well-respected physicists deny that it is even a theory within the definition of the scientific method and refer to it as a 'philosophy' instead, placing it on the boundary between physics and metaphysics. This has not stopped string theorists from continuing to develop the theory, especially since there may be some tests within the reach of current technology of revised versions of string theory as described below.

14.2 M-theory

Unfortunately, further work on string theory in the 1990s revealed that there were not one, but five different entirely self-consistent versions of $N = 10$ string theory—not a very encouraging state of affairs for those who want to develop a single unified TOE. Then, in 1995, Edward Whitten noticed that all five flavors of the TOE could be unified into one theory, at the cost of introducing yet another dimension in which the strings were seen to actually be two-dimensional membranes, or *branes* for short. String theory had morphed into something that Whitten called *M-theory*. A remarkable feature of 11-dimensional M-theory is that the extra dimension could actually be large. If that is the case, then why can we not perceive it? The idea here is that the strings corresponding to (almost) all the particles in our Universe have their ends firmly attached to a brane within a higher-dimensional hyperspace called the *bulk* (see figure 14.2). As a result we, being made up of these particles, cannot access the larger dimension outside our home brane. One image often used is that of a loaf of bread making up the bulk. The entire Universe we are familiar with is then only a single slice of the loaf. It is then entirely possible that other Universes could exist on other branes, possibly as close as millimeters away from us, which we cannot perceive.

One of the more intriguing features of this *brane cosmology* is that it has the possibility to account for the extreme weakness of gravity, which is very difficult to understand in more conventional theories. The idea is that the graviton string, instead of having its ends firmly tied to our brane, is actually a closed loop that can roam freely throughout the bulk. This *leakage* of gravity reduces its apparent strength in our own Universe. A corollary is that gravity might appear to be stronger at very small (possibly subatomic) distances where less of the gravitational force might have had an opportunity to leak out. Experiments are being carried out to test

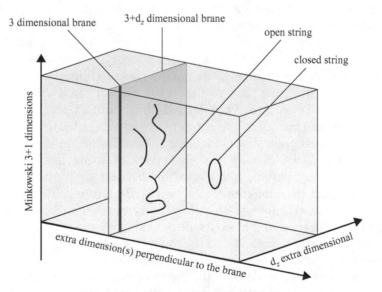

Figure 14.2. A diagram illustrating M-theory.

this hypothesis but so far the results have been inconclusive at best. Leakage of gravity into the bulk also suggests the potential for us to communicate with a nearby Universe if gravitational waves generated there could be detected. Another application is to the theory of dark energy. Calculations of dark energy made under the assumption that it is due to Einstein's cosmological constant have concluded that its strength should be something like 10^{122} times (!) greater than its observed value. This has been called the most spectacularly wrong prediction ever made. The brane cosmology reduces or eliminates this discrepancy due to the fact that gravity can leak into another dimension. Finally, there is even a potential application to the cause of the Big Bang itself, which might have resulted from the collisions of two branes moving in the bulk! These are, of course, highly speculative (some would say metaphysical) suggestions and will remain so unless and until a firm experimental footing for M-theory can be established. It appears possible that versions of the theory containing large extra dimensions might make predictions that could be tested in the future, even at LHC energies.

14.3 The multiverse

The *brane-world* scenario discussed above naturally leads to the idea that there might be many other Universes in addition to our own—the so-called *multiverse hypothesis*. As with other such 'wild' suggestions, the physics community is divided on this issue, with some scientists arguing that, in the absence of compelling experimental evidence for M-theory, the multiverse is at best metaphysical or possibly even pseudoscientific. The concept is appealing, however, for a number of reasons. From the point of view of the string-theory landscape, it suggests that the many possible Universes from different compactifications might really exist. The fundamental constants of physics, such as Newton's gravitational constant or the charge on the electron or the Planck constant, would have different values in these different Universes. Now, it turns out that the existence of life as we know it is extraordinarily sensitive to the values of these constants. This leads to the so-called *anthropic principle*, which suggests that our Universe is the way it is because we are here to observe it. Other Universes with different parameters may well exist, but they would remain sterile unless their properties would also allow for the existence of observers. Steven Weinberg has used this idea to account for the very small but non-zero value of the cosmological constant. As mentioned above, various calculations suggest values for this constant up to 10^{122} times its observed value. However, it turns out that Universes having such large cosmological constants would expand very rapidly, before galaxies or any other structures could ever form. Only those Universes that have values for this constant near to what we actually observe could possibly support life and therefore observers. The multiverse together with the anthropic principle may also have an application to the 'why' of the Big Bang. As mentioned above, it has been suggested that new Universes may continually be being created by collisions of branes in the bulk. However, only a certain subset of these new Universes could ever develop into something we might recognize. The explanation goes back to a suggestion made by Edward Tryon in 1973. He pointed out that gravitational energy is negative while other forms of matter and energy correspond to positive values. If gravity and positive energy exactly balance then the total energy is zero and this turns out to be the case if and only if $\Omega = 1$ as in our own Universe. Now consider what happens when we substitute this into the Heisenberg uncertainty principle $\Delta t = h/(4\pi\Delta E)$. This is written in a slightly different way than before to emphasize Tryon's point which was that, if we substitute $\Delta E = 0$ into this equation, the corresponding value of Δt is infinite. In his words, the Universe could then be a 'quantum fluctuation of the vacuum' which lives forever (or at least for a very long time). An entire Universe could appear out of nothing, to quote Steven Hawking. Alan Guth referred to this as the 'ultimate free lunch'. (It is worth pointing out, however, that the vacuum state is not really 'nothing' in the sense we usually think of this term, since it is actually a complex soup of virtual particles.) In any case, to return to the discussion above, Universes formed from the collisions of branes would not survive for any length of time unless they happened to have $\Omega = 1$.

For truly excellent videos of the concepts discussed in this chapter, see Brian Greene's PBS-NOVA series *The Elegant Universe* and *The Fabric of the Cosmos*.

Elementary Cosmology: From Aristotle's Universe
to the Big Bang and Beyond

James J Kolata

Chapter 15

Black holes and wormholes

15.1 The life of the Sun

Stars form in the collapse of interstellar dust and gas clouds under their mutual gravitational attraction. This process is going on even now, as witnessed, for example, by star-forming regions observed in the Orion nebula. This object is the remnant of an ancient supernova explosion and the stars being born from it contain elements like carbon, nitrogen, and oxygen formed during the lifetime of the progenitor star. Thus they are similar to our Sun, which also was formed, 5 billion years ago, from just such a *contaminated* cloud of dust and gas that collapsed and heated up until nuclear fusion reactions of hydrogen began to occur. The energy streaming from its center eventually stopped the collapse of the proto-Sun, which then reached *hydrostatic equilibrium*. The subsequent history of a star depends critically upon its mass. Very massive stars need to be very hot in order to reach hydrostatic equilibrium and nuclear fusion occurs much faster in them because of this. Eventually the hydrogen gas in their cores is entirely fused into helium, within only about 50 million years for a 25 solar mass star. While the same thing will happen to our Sun, its expected lifetime is about 10 billion years. At this point, hydrogen-fusion energy disappears and the star again begins to collapse under gravitational attraction until the temperature at its core rises sufficiently for helium to fuse. (Helium fuses at a higher temperature than hydrogen because it has two electrical charges rather than one, so the electrostatic repulsion that must be overcome is greater.) When helium fusion begins, the star again reaches hydrostatic equilibrium but its outer atmosphere expands and cools thus forming a *red-giant* star. In the case of the Sun, its surface will then be somewhere between the orbits of the Earth and Mars. The red-giant phase lasts for only about 100 million years until the helium fuel is used up. This time, however, the gravitational mass is too small to ignite fusion of heavier elements such as the carbon formed from helium fusion. The

Sun will continue to contract and heat up until the atoms in its core are packed so tightly together that the repulsion of their electron clouds stops the collapse. In this process its outer layers are completely ejected, forming a *planetary nebula* with an extremely hot *white dwarf* star at its center. The radius of the Sun will go from the present 7×10^5 km to only 1×10^4 km and its density will then be about 350 000 times that of water. Over many billions of years, the white dwarf will cool and eventually disappear from view. This is the fate awaiting stars with masses less than about 3.5 times the mass of the Sun.

15.2 The life of massive stars

The 25 solar mass star mentioned above will also go through the helium-fusing stage, forming a red super-giant star, but will run out of helium fuel in its core in only about 1 million years. Big stars live fast and die young! It also has a very different fate in store for it, since such stars have enough mass to ignite fusion of successively heavier elements. In a relatively short time, it will fuse first carbon, then magnesium, then silicon, and finally sulfur to form iron. According to calculations of this process, the resulting object has an iron core with layers of successively lighter elements arranged in an onion-like structure as shown in figure 9.2.

However, the star now has a serious problem. According to well-understood laws of nuclear physics, fusion of iron does not generate energy. Instead, it acts like a nuclear refrigerator instead of a nuclear furnace. The star, which once was stabilized by the energy of fusion, now has no way to prevent its collapse under the influence of gravity and it first implodes and then bounces back and is destroyed in a supernova explosion. The elements 'cooked up' in its various fusion stages are ejected into the surrounding interstellar medium, perhaps to be incorporated into later generations of stars and, in the case of the Sun, into the bodies of plants and animals on the Earth. Observations of supernova remnants have detected all these elements, including iron. However, at least a portion of the inner core of the supernova may escape this dispersion, resulting in the formation of more exotic objects. Calculations have also shown that the elements heavier than iron are formed in the very energetic supernova explosion itself. The exact details of this process are the subject of ongoing research in nuclear astrophysics.

15.3 Neutron stars

Compression of the matter in the supernova core can push through the repulsion of the electron clouds, forming a state of matter in which the atomic nuclei are packed tightly together and the electrons are pushed back into the protons to form a *neutron star*. This object has a density of about 5×10^{14} times that of water, corresponding to the entire mass of the Sun compressed into only a 10 km radius! Because of the physical law of conservation of angular momentum, the star's rate of spin increases dramatically during the collapse, up to as much as 40 000 revolutions per minute. Also, if like the Sun the progenitor star had a significant magnetic field, the compression will cause it to increase, typically to something like a magnetic field (at its surface) that is a trillion times that of the Earth. However, neutron stars having hundreds of times larger magnetic fields, called *magnetars*, have been detected. Their fields are so powerful that they could kill a person from 1000 km away by warping the atoms in living flesh! However, they pose a threat at much greater distances than that because of the powerful flares they eject. Due to these magnetic fields neutron stars are relatively easy to detect, because they form what are known as *pulsars*. The first of these was discovered in July 1967 by Jocelyn Bell. Using a radiotelescope, she found that the signal from a peculiar object pulsed on and off in a very regular pattern, at a rate of approximately one pulse per second. It was initially called LGM-1 (for 'little green men') because of the suggestion that the signal might be an attempt at communication from intelligent inhabitants of a distant star. However, it was soon realized that this was not the case and the explanation for the regular pulsation was more prosaic but almost as interesting. Flares emitted from the surface of a neutron star are channeled through its magnetic poles, just as the charged particles emitted by our Sun during a solar flare are directed by the Earth's magnetic field to form the *aurora borealis* and the *aurora australis* when they strike the Earth's atmosphere. In the case of neutron stars, these flares emit radio waves along the magnetic poles. If, as in the case of the Earth, the magnetic poles are not aligned with the geometric poles of rotation, then the corresponding radio emissions rotate in space at the frequency of rotation of the neutron star. If the Earth happens to be in a location such that these radio waves sweep past it, then the result is a pulsing emission much like the flashing light you see from a lighthouse (figure 15.1).

This discovery was awarded the 1974 Nobel Prize in physics but Jocelyn Bell, who was a postdoctoral scholar in the laboratory of Martin Ryle and Anthony Hewish, controversially did not share in the prize. Pulsars are very accurate clocks. In fact, the most accurate clock in the world (at the time) was the *pulsar clock* installed in Gdansk, Poland in 2011. However, they do slow down over long periods of time due to the emission of energy and this slowdown has also been detected.

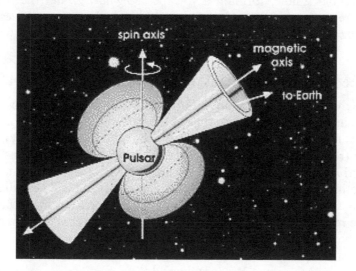

Figure 15.1. The 'lighthouse effect' for pulsars. Image courtesy of NASA/GFSC.

15.4 Black holes

If the mass of a neutron star exceeds about 1.5–3.0 solar masses (the exact value is at present not well known), then its gravitational pressure will be so great that even neutrons cannot withstand it. The surface of the neutrons is breached and the collapse continues catastrophically forming a *black hole* which is a region of space-time from which gravity prevents anything, even light, from escaping. The possibility that this might happen was discussed by John Michell in 1783 and at about the same time by Pierre Laplace, but the first solution of Einstein's general theory of relativity showing that these objects can exist was published by Karl Schwarzschild in 1916. The principle can best be understood in analogy with what happens to objects thrown upward from the surface of the Earth. As we all know, a baseball thrown directly upward will slow down under the influence of gravity, eventually stop, and then fall back to Earth. However, a rocket ship traveling fast enough (at the *escape velocity* of about 25 000 miles per hour) will break free from the Earth's gravity and leave without ever falling back to Earth. It turns out that the square of the escape velocity from the surface of an object is dependent only on the ratio of the mass of the object to its radius ($v^2 = 2GM/R$, where G is Newton's gravitational constant). If the mass M is large enough or the radius R small enough (or both), then the speed v will equal the speed of light and not even light can escape from the surface of the object. Since the special theory of relativity states that nothing can travel through space faster than light, nothing can escape the gravitational pull at the 'surface' of a black hole. Schwarzschild calculated the so-called *Schwarzschild radius* of a non-rotating black hole of mass M ($R_s = 2GM/c^2$, where c is the speed of light) which depends only on the mass of the object. For example, R_s for the Earth turns out to be only 9 mm. If some process were to compress the Earth into a ball of this radius, it would become a black hole. The more massive the black hole, the bigger it is, but the Schwarzschild radius is not like the radius of a ball bearing. All of the matter within R_s falls into the center of the black hole, forming a so-called *singularity*, and the Schwarzschild radius only represents the boundary from within which nothing can escape, called the *event horizon*.

15.5 Some properties of black holes

Since no light, and therefore no information, can come from inside R_s, only quantities that are *conserved* according to the laws of physics can remain after the black hole is formed. In particular, these are the mass, charge (if any) and angular momentum (rotation) of the matter within its surface. This is known as the *no-hair theorem*, and it implies that a black hole cannot have a magnetic field. If a neutron star prior to its collapse possessed a magnet field, that field will be radiated away as electromagnetic radiation during the collapse. That makes it quite difficult to detect a black hole resulting from a supernova explosion, unlike the case of a neutron star. Only a few have been detected, and they are in binary systems where matter from a neighboring star is pulled into a black hole generating a lot of energy. An example is the object known as Cygnus X-1 (see figure 15.2).

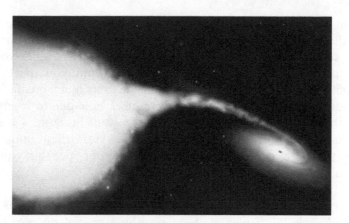

Figure 15.2. An artist's impression of the Cygnus X-1 binary system. Image credit: ESA/Hubble. Image courtesy of NRAO/AUI.

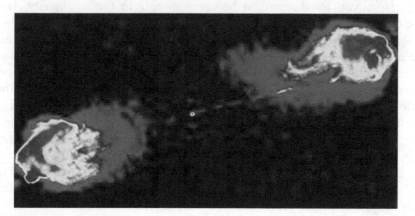

Figure 15.3. The galaxy Cygnus A with a supermassive black hole in its center. Image courtesy of NRAO/AUI.

However, there is another way in which black holes can form. The density of stars in the center of a galaxy is very high and they can collide and merge, eventually forming a *supermassive* black hole. In the case of the Milky Way, observations have shown that there exists a black hole at its center with a mass about 4 million times that of the Sun. Analysis of the motion of stars around this object, known as Sagittarius A*, have confirmed its mass and radius demonstrating that it must be a black hole. Supermassive black holes having masses up to billions of times that of the Sun have been detected in other galaxies from the energetic jets that emerge from them as stars fall into the black hole. An example is the galaxy known as Cygnus A shown in figure 15.3. It is thought that every galaxy may well have a supermassive black hole at its center.

The (Schwarzschild) radius of a black hole increases when matter falls into it to account for its greater mass. It also turns out that the *gravitational tidal effects* at the event horizon become weaker as the mass and radius of the black hole increases. To understand these tidal effects, think about the effect of gravity on your body as you stand on the surface of the Earth. Since gravity is slightly weaker at your head than at your feet, there is a very small net force tending to pull you apart. This effect is greatly magnified near a solar-mass black hole due to its much greater gravitational attraction, but it is smaller for a supermassive black hole due to its enormous size. This means that you could in principle cross the event horizon of a supermassive black hole, such as the one at the center of our Galaxy, without immediately being torn apart by tidal forces and you might in fact not even notice that you had actually passed the point of no return!

An interesting sidelight of falling into a black hole concerns the nature of this event as observed by you (as you fall into the black hole) and someone at a distance from the black hole. You would observe nothing special on crossing the event horizon, and it would take only a short while (as measured by your clocks) for you to fall into the *singularity* at its center. There is no way to avoid this. Once you cross the event horizon, you are doomed to be compressed into the singularity. However, because of *gravitational time dilation* an observer at a distance from the black hole would see you approach the event horizon but never cross it! This is one of the strange effects of the general theory of relativity.

The rotation rate of a black hole has a natural limit, when the event horizon is rotating at the speed of light. This means that the larger (i.e. more massive) the black hole, the slower its maximum rate of spin. It turns out that the event horizon of a non-rotating black hole is spherical, which prevents us from seeing the singularity at its center. This has become known as *cosmic censorship*. However, the event horizon of a rapidly spinning black hole is deformed and bulges at the equator (just like the Earth). This brings up the interesting question if the deformation could ever be great enough that we could directly observe the *naked singularity* at its center. This question is not answered yet. It was the subject of a famous bet between Steven Hawking and Kip Thorne. The rotation of a black hole is interesting for another reason. A solution of Einstein's general relativity equations for this case (the *Kerr solution*) reveals that the rotation of a black hole causes a 'swirling' of space-time outside of it—i.e. outside the event horizon. Because of this, it turns out to be possible to extract a really huge amount of rotational energy from a spinning black hole (thus slowing down its rate of spin). This is thought to be the source of the energy that powers the immense jets of matter and energy emerging from galactic cores containing supermassive black holes. It is also presumed to be the source of the energy that powers *quasars*, distant *quasi-stellar* objects that look like points of light but have very large red shifts so they are very far away. Recently, galaxies surrounding some of these quasars have been observed. The energy output of these objects is variable on the time scale of days or shorter, so they must be very compact. Only emission from supermassive black holes can account for these properties. (Quasars are all very distant from us, so it is thought that they are the end result of the evolution of galaxies formed when the Universe was much younger—possibly the first galaxies to form.)

15.6 The thermodynamics of black holes

Steven Hawking deduced some fundamental and important properties of the energetics of black holes from a consideration of a fundamental principle of physics, the second law of thermodynamics. In order to discuss this, we need to introduce the concept of *entropy*, which is a measure of the *disorder* of a system. As an example, consider the number of different sequences that can result from flipping n coins. For $n = 3$, there are 8 possibilities, for $n = 4$ there are 16 possibilities, and for $n = 100$ there are 2^{100} possibilities. The entropy for a random toss of 100 coins is $\log(2^{100}) = 30.1$, where 'log' indicates the logarithm to the base 10. In this example, $10^{30.1} = 2^{100}$. Logarithms are used because one rapidly gets very big numbers for most systems and large entropy means a large degree of disorder. The second law of thermodynamics states that *the entropy of an isolated system always increases*. That does not mean that disorder is inevitable. For example, I could choose to arrange the sequences of coin tosses in some predetermined order, thereby lowering its entropy. However, it takes energy (mental and physical) for me to do this and the second law says that the total entropy of the system (me plus the coins) increases in this ordering process. My use of internal and external energy increases my entropy more than enough to compensate for the decrease in the entropy of the coins. Viewed in this way, the order of living beings on the Earth (as only one example) is made possible by energy from the Sun, which in the process increases the Sun's entropy. The second law is an extremely powerful tool and it is closely bound up with the nature of time. For example, consider watching a movie of a glass bottle shattering under the impact of a rock. If you run the movie backwards, it looks strange. Why? Because the disordered glass fragments rearrange themselves into an ordered bottle—and this never happens by itself. The direction of time flow is related to the increase in entropy. Another thing to consider: no physical principle other than the second law prevents all the air molecules in a room from gathering by chance in one corner—which would be a disaster for us, of course. In fact, this *can* happen by chance, but the probability it *will* happen is vanishingly small. The second law deals with the probabilities of such rare events. As such, it is different from physical laws such as Newton's laws of motion because these make absolute predictions. The second law is more akin to *probabilistic* theories such as quantum mechanics.

Jacob Bekenstein was one of the first to notice that there is a deep connection between the entropy of a black hole and the area of its event horizon. In fact, he showed that *the entropy of a black hole is equal to the ratio of this area to the square of the Planck distance*. For a 10 solar mass black hole, $R_s = 30$ km, $A = 2.8 \times 10^9$ m^2, and the entropy is 10^{79} (not 79, the log has already been taken)! This is an enormous amount of disorder for an object that is supposed to be simple. Remember: 'a black hole has no hair'. Initially this was viewed as a problem, which was one of the things that retarded the study of black-hole thermodynamics. In retrospect, it is clear that this result is mandated by the second law. For example, we could take all the gas molecules in a room, which have a very big entropy because of their random motions, and dump them into a black hole. In the process, a lot of entropy disappears from the neighborhood because all we can know about the air molecules within the black hole is their total mass (and electric charge and angular momentum if they had any). Their random motions are not visible anymore. The second law says that the total entropy of the Universe must increase, however, so the entropy of the black hole has to go up. In fact, it is now understood that *the entropy of a black hole is a measure of the number of different ways we can form it*—which is very large. This disorder must reside inside the black hole and according to Bekenstein's theorem the area of the event horizon, which increases in size as more mass is added, is a measure of the disorder contained within the black hole.

15.7 Hawking radiation

We now come to Steven Hawking's insight. In 1974, he put it all together when he showed that not only was the area of its event horizon related to the entropy of a black hole, but also the surface gravity at R_s was a measure of an *effective temperature* of the black hole. In the process, he showed that black holes can radiate not only photons but also particles of many kinds, and the spectrum of this radiation was given by the Planck formula. This is now known as *Hawking radiation*. How does it work? The process goes something like this: a quantum fluctuation produces, e.g., an electron–antielectron pair near the event horizon, which is then torn apart by the tidal gravity of the black hole. This makes the 'virtual' particles real by taking energy from the gravitational field. One of the pair then drops into the hole and the other radiates into space. Of course, the same thing can happen with virtual photons, so the hole also radiates electromagnetic radiation. In Hawking's words: 'black holes ain't black'. What are the properties of this radiation? First of all, Hawking showed that *the temperature of a black hole is inversely proportional to its mass*. The temperature of a 4 solar mass black hole, for example, is 1.5×10^{-8} K. This is much less than the temperature of the CMB, which means that in fact it is a net absorber of radiation from the CMB. All 'normal' black holes are heavier and therefore even colder than this. However, there may be an exception to the rule. Hawking speculated that *primordial black holes* might have been made in the Big Bang and survived to the present. Hawking radiation implies a loss of energy (and therefore mass) from the black hole. But a smaller black hole has a higher temperature and the emission of radiation goes as the fourth power of the temperature, so the hole heats up as it shrinks and therefore loses mass even faster. How long does it take for the black hole to completely evaporate? It can be shown that the lifetime of a black hole (neglecting absorption of the CMB) is proportional to the cube of its mass. For a 10 solar mass black hole, it works out to 2×10^{78} s. The present age of the Universe is about 5×10^{17} s so we do not have to worry about evaporation of 'normal' black holes. However, one can ask for the mass of a black hole that lives for the current age of the Universe. The answer turns out to be 1.3×10^{11} kg, corresponding to a radius of about 20% of that of a proton. Now, the Hawking process is a prescription for a *thermal runaway* in which most of the energy is emitted close to the end leading to an explosion. Black holes of this mass should be exploding right now. Events such as this are being looked for, but so far none have been seen. There is also a problem called the *black hole information paradox* associated with evaporation of a black hole. As mentioned above, almost all the information associated with matter falling into a black hole disappears. This was not a problem at first when it was thought that the lifetime of a black hole is infinite. The 'lost' information resides inside the black hole. However, the Hawking process does not take any information from the inside of the black hole, so where does this information go when the black hole evaporates? Loss of information turns out to be a serious problem for physics and the community is currently divided on whether the information is actually lost.

15.8 The singularity at the center of a black hole

All the matter falling through the event horizon of a black hole should form a gravitational singularity at its center. Classically, this would be a one-dimensional point object where all the mass is concentrated into an infinitely small space and space-time has infinite curvature. All physical laws break down in such an extreme environment. Quantum mechanics tells us that this could never happen since the size of the singularity will be the Planck length. Nevertheless, it takes a quantum theory of gravity (currently unavailable) in order to understand the nature of the singularity. Even so, various speculations about black holes and singularities have been put forward. As with many of the speculations surrounding string theory, these are on the boundary between physics and metaphysics. Among them is the *white hole*, a hypothetical region of space-time from which matter and energy emerge from a singularity. In this sense, it is the reverse of a black hole, and possibly the place where the information lost in a black hole reappears. Such a solution of the general theory of relativity has been found, but there are no known physical processes by which a white hole could form. More intriguing is the *Einstein–Rosen bridge* or *wormhole*, which is like a tunnel with its two ends at separate points in space-time (see figure 15.4). If such a thing existed, it could form a shortcut from one point in space to another very far distant point, allowing for travel at speeds much faster than light.

As an alternative, it could also make time travel possible or even allow travel into another brane in the multiverse. However, it has been shown that wormholes that could actually be traversed from one end to another would only be possible if they could be stabilized by an exotic form of matter with negative energy density to open up the singularity at the center. Physicists such as Steven Hawking and Kip Thorne have argued that properties of quantum mechanics might allow for such things. These ideas play a big role in the movie *Interstellar* for which Kip Thorne was creative director and executive producer.

Another possibility under discussion is that the quantum foam at the earliest stages of the Big Bang actually was formed from tiny wormholes that continually appear and disappear according to the laws of quantum mechanics. Finally, there is the possibility that our entire Universe began from a singularity and is itself a black hole! This, as might be expected, is quite controversial, and Steven Hawking has put forward a *no-boundary* hypothesis in which the Universe evolved in a smooth way

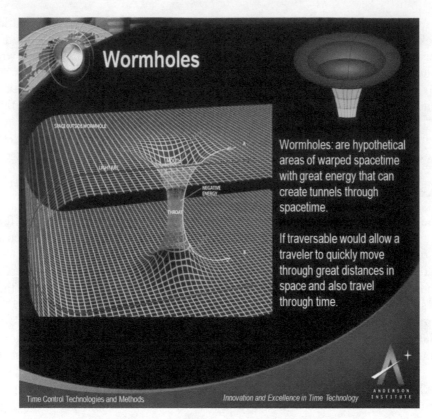

Figure 15.4. A description of a wormhole. Image courtesy of The Anderson Institute.

from a single point which was not actually a singularity, in much the same way as the geometric poles of the Earth are unique points that are not actually singularities. Imagine tracing the intersection of the Earth with a plane moving downward from the North Pole toward the Equator. The result would be a single, one-dimensional point that then expanded smoothly into a circle. At the Equator, the circle would reach its maximum diameter and begin to shrink until it again became a point as the plane reached the South Pole. This is in fact part of the content of the no-boundary hypothesis, which predicts that the Universe will eventually collapse into a point, after a very long time.

Chapter 16

Reading list

Further reading

Thorne K S 1995 Black Holes and Time Warps: Einstein's Outrageous Legacy (Commonwealth Fund Book Program) (W W Norton & Company) ISBN-13: 9780393312768 ISBN-10: 0393312763

Barrow J D 1988 The World Within the World (Oxford: Oxford University Press) ISBN-10: 0198519796 ISBN-13: 978-0198519799

Chown M 2003 The Universe Next Door: The Making of Tomorrow's Science (Oxford: Oxford University Press) ISBN-13: 9780195168846 ISBN-10: 0195168844

Gribbin J 2001 Hyperspace: The Universe and Its Mysteries (DK Adult) ISBN-13: 9780789478382 ISBN-10: 0789478382

Steinhardt P J and Turok N 2007 Endless Universe: Beyond the Big Bang (New York: Doubleday) ISBN-13: 9780385509640 ISBN-10: 0385509642

Overbye D 1999 Lonely Hearts of the Cosmos: The Story of the Scientific Quest for the Secret of the Universe (Little, Brown & Company) ISBN-13: 9780316648967 ISBN-10: 0316648965

Davies P 1996 About Time: Einstein's Unfinished Revolution (Simon & Schuster) ISBN-13: 9780684818221 ISBN-10: 0684818221

Halpern P 2005 The Great Beyond: Higher Dimensions, Parallel Universes and the Extraordinary Search for a Theory of Everything (New York: Wiley) ISBN-13: 9780471741497 ISBN-10: 0471741493

Randall L 2006 Warped Passages: Unravelling the Mysteries of the Universe's Hidden Dimensions (Harper Perennial) ISBN-13: 9780060531096 ISBN-10: 0060531096

Barrow J D 1997 The Origin Of The Universe: Science Masters Series (Basic Books) ISBN-13: 9780465053148 ISBN-10: 0465053149

Narlikar J and Burbidge G 2008 Facts and Speculations in Cosmology (Cambridge: Cambridge University Press) ISBN-13: 9780521865043 ISBN-10: 0521865042

Hazen R M 1997 Why Aren't Black Holes Black? (Anchor) ISBN-13: 9780385480147 ISBN-10: 0385480148

Kaku M 2006 Parallel Worlds: A Journey Through Creation, Higher Dimensions, and the Future of the Cosmos (Anchor) ISBN-13: 9781400033720 ISBN-10: 1400033721

Vilenkin A 2006 Many Worlds in One: The Search for Other Universes (Hill and Wang) ISBN-13: 9780809095230 ISBN-10: 0809095238

Rees M 2003 Our Cosmic Habitat (Princeton, NJ: Princeton University Press) ISBN-13: 9780691114774 ISBN-10: 0691114773

Krauss L M 200 Quintessence: The Mystery of Missing Mass in the Universe (Basic Books) ISBN-13: 9780465037414 ISBN-10: 0465037410

Nicolson I 2007 Dark Side of the Universe: Dark Matter, Dark Energy, and the Fate of the Cosmos (Baltimore: Johns Hopkins University Press) ISBN-13: 9780801885921 ISBN-10: 0801885922

Barrow J D 2002 The Book of Nothing: Vacuums, Voids, and the Latest Ideas about the Origins of the Universe (Vintage) ISBN-13: 9780375726095 ISBN-10: 0375726098

Pickover C A 2001 Surfing Through Hyperspace: Understanding Higher Universes in Six Easy Lessons (Oxford: Oxford University Press) ISBN-13: 9780195142419 ISBN-10: 0195142411

Wilczek F and Devine B 1989 Longing for the Harmonies: Themes and Variations from Modern Physics (W W Norton & Company) ISBN-13: 9780393305968 ISBN-10: 0393305961

Polkinghorne J C 1990 Quantum World (Penguin Books Ltd) ISBN-13: 9780140134926 ISBN-10: 0140134921

Davies P and Gribbin J 2007 The Matter Myth: Dramatic Discoveries that Challenge Our Understanding of Physical Reality (Simon & Schuster) ISBN-13: 9780743290913 ISBN-10: 0743290917

Chown M 2010 Afterglow of Creation: Decoding the Message from the Beginning of Time (Faber & Faber) ISBN-13: 9780571250592 ISBN-10: 0571250599

Kolb R 1997 Blind Watchers Of The Sky: The People And Ideas That Shaped Our View Of The Universe (Basic Books) ISBN-13: 9780201154962 ISBN-10: 020115496X

Lederman L M and Hill C T 2008 Symmetry and the Beautiful Universe (Prometheus Books) ISBN-13: 9781591025757 ISBN-10: 1591025753

Livio M 2000 The Accelerating Universe: Infinite Expansion, the Cosmological Constant, and the Beauty of the Cosmos (New York: Wiley) ISBN-13: 9780471399766 ISBN-10: 0471399760

Lang K 2014 Parting the Cosmic Veil (Berlin: Springer) ISBN-13: 9781493900930 ISBN-10: 1493900935

Begelman M and Rees M 2009 Gravity's Fatal Attraction: Black Holes in the Universe (Cambridge: Cambridge University Press) ISBN-13: 9780521717939 ISBN-10: 0521717930

Silk J 2008 The Infinite Cosmos: Questions from the Frontiers of Cosmology (Oxford: Oxford University Press) ISBN-13: 9780199533619 ISBN-10: 019953361X

Gribbin J 1999 The Search For Superstrings, Symmetry, And The Theory Of Everything (Little, Brown and Co) ISBN-13: 9780316329750 ISBN-10: 0316329754

Chown M 2001 The Magic Furnace: The Search for the Origins of Atoms (Oxford: Oxford University Press) ISBN-13: 9780195143058 ISBN-10: 0195143051

Kaku M and Thompson J T 1995 Beyond Einstein: The Cosmic Quest for the Theory of the Universe (Anchor) ISBN-13: 9780385477819 ISBN-10: 0385477813

Zirker J B 2001 Journey from the Center of the Sun (Princeton, NJ: Princeton University Press) ISBN-13: 9780691057811 ISBN-10: 0691057818

Adams F C and Laughlin G 2002 The Five Ages of the Universe: Inside the Physics of Eternity (New York: Free Press) ISBN-13: 9780684865768 ISBN-10: 0684865769

Hogan C J 1999 The Little Book of the Big Bang: A Cosmic Primer (Copernicus) ISBN-13: 9780387983851 ISBN-10: 0387983856

Greene B 2005 The Fabric of the Cosmos: Space, Time, and the Texture of Reality (Vintage) ISBN-13: 9780375727207 ISBN-10: 0375727205

Chown M 2006 The Quantum Zoo: A Tourist's Guide to the Never-Ending Universe (J H Press)

Gribbin J 2000 The Birth of Time: How Astronomers Measured the Age of the Universe (New Haven, CT: Yale University Press) ISBN-13: 9780300083460 ISBN-10: 0300083467

Webb S 2011 Out of this World: Colliding Universes, Branes, Strings, and Other Wild Ideas of Modern Physics (Copernicus) ISBN-13: 9781441918529 ISBN-10: 1441918523

De Grasse Tyson N 2014 Death by Black Hole: And Other Cosmic Quandaries (W W Norton & Company) ISBN-13: 9780393350388 ISBN-10: 039335038X

Hawking S W 2001 The Universe in a Nutshell (Bantam) ISBN-13: 9780553802023 ISBN-10: 055380202X

Hawking S W 1996 The Illustrated Brief History of Time, Updated and Expanded Edition (Bantam) ISBN-13: 9780553103748 ISBN-10: 0553103741

Levin F 2006 Calibrating the Cosmos: How Cosmology Explains Our Big Bang Universe (Berlin: Springer) ISBN-13: 9780387307787 ISBN-10: 0387307788

Calder N 1979 Einstein's Universe (Viking) ISBN-13: 9780670290765 ISBN-10: 0670290769

De Grasse Tyson N and Goldsmith D 2014 Origins: Fourteen Billion Years of Cosmic Evolution (W W Norton & Company) ISBN-13: 9780393350395 ISBN-10: 0393350398

Stenger V J 2000 Timeless Reality: Symmetry, Simplicity, and Multiple Universes (Prometheus Books) ISBN-13: 9781573928595 ISBN-10: 1573928593

Wright A D 1989 At the Edge of the Universe (Ellis Horwood Ltd) ISBN-13: 9780470212394 ISBN-10: 047021239X

Fairall A 2001 Cosmology Revealed (Berlin: Springer) ISBN-13: 9781852333225 ISBN-10: 1852333227

Gribbin J 1999 In Search of the Edge of Time: Black Holes, White Holes, Wormholes (Penguin Books) ISBN-13: 9780140248142 ISBN-10: 0140248145

Goldsmith D 1997 Einstein's Greatest Blunder?: The Cosmological Constant and Other Fudge Factors in the Physics of the Universe (Cambridge, MA: Harvard University Press) ISBN-13: 9780674242425 ISBN-10: 0674242424

Davies P 1995 The Edge of Infinity: Beyond the Black Hole (Penguin Science) ISBN-13: 9780140231946 ISBN-10: 0140231943

Barnett R M, Muehry H and Quinn H R 2002 The Charm of Strange Quarks: Mysteries and Revolutions of Particle Physics (Berlin: Springer) ISBN-13: 9780387988979 ISBN-10: 0387988971

Kane G 2001 Supersymmetry: Unveiling The Ultimate Laws Of Nature (Basic Books) ISBN-13: 9780738204895 ISBN-10: 0738204897

Bernstein J 1998 An Introduction to Cosmology (Pearson P T R) ISBN-13: 9780139055485 ISBN-10: 0139055487

Hawley J F and Holcomb K A 2005 Foundations of Modern Cosmology (Oxford: Oxford University Press) ISBN-13: 9780198530961 ISBN-10: 019853096X

Guth A 1998 The Inflationary Universe (Basic Books) ISBN-13: 9780201328400 ISBN-10: 0201328402

Greene B 2003 The Elegant Universe: Superstrings, Hidden Dimensions, and the Quest for the Ultimate Theory (W W Norton & Company) ISBN-13: 9780393058581 ISBN-10: 0393058581

Green B 2011 The Hidden Reality: Parallel Universes and the Deep Laws of the Cosmos (Vintage) ISBN-13: 9780307278128 ISBN-10: 0307278123

Weinberg S 1993 The First Three Minutes: A Modern View Of The Origin Of The Universe (Basic Books) ISBN-13: 9780465024377 ISBN-10: 0465024378

Davies P 1997 The Last Three Minutes: Conjectures About The Ultimate Fate Of The Universe (Basic Books) ISBN-13: 9780465038510 ISBN-10: 0465038514

Kirshner R P 2004 The Extravagant Universe: Exploding Stars, Dark Energy, and the Accelerating Cosmos (Princeton, NJ: Princeton University Press) ISBN-13: 9780691117423 ISBN-10: 069111742X

Rees M 2001 Just Six Numbers: The Deep Forces That Shape The Universe (Basic Books) ISBN-13: 9780465036738 ISBN-10: 0465036732

Schumm B A 2004 Deep Down Things: The Breathtaking Beauty of Particle Physics (Johns Hopkins Univeristy Press) ISBN-13: 9780801879715 ISBN-10: 080187971X

Stewart I 2002 Flatterland: Like Flatland, Only More So (Basic Books) ISBN-13: 9780738206752 ISBN-10: 073820675X

Gamow G 2001 The New World of Mr Tompkins: George Gamow's Classic Mr Tompkins in Paperback (Cambridge: Cambridge University Press) ISBN-13: 9780521639927 ISBN-10: 0521639921

Silk J 2000 The Big Bang Third Edition (Times Books) ISBN-13: 9780805072563 ISBN-10: 080507256X

Singh S 2005 Big Bang: The Origin of the Universe (Harper Perennial) ISBN-13: 9780007162215 ISBN-10: 0007162219

Ferreira P 2007 The State of the Universe: A Primer in Modern Cosmology (Phoenix) ISBN-13: 9780753822562 ISBN-10: 0753822563

Seife C 2004 Alpha and Omega: The Search for the Beginning and End of the Universe (Penguin Books) ISBN-13: 9780142004463 ISBN-10: 0142004464

Kidger M 2007 Cosmological Enigmas: Pulsars, Quasars, and Other Deep-Space Questions (Johns Hopkins University Press) ISBN-13: 9780801884603 ISBN-10: 0801884608

Allen H W G 2001 The New Cosmology (Perspective Books) ISBN-13: 9780962455544 ISBN-10: 0962455547

Mallary D M 2004 Our Improbable Universe: A Physicist Considers How We Got Here (Thunder's Mouth Press) ISBN-13: 9781568583013 ISBN-10: 156858301X

Silk J 2005 On the Shores of the Unknown: A Short History of the Universe (Cambridge Univeristy Press) ISBN-13: 9780521836272 ISBN-10: 0521836271

Rees M 1998 Before The Beginning: Our Universe And Others (Basic Books) ISBN-13: 9780738200330 ISBN-10: 0738200336

Silk J I 1997 Cosmic Enigmas (Masters of Modern Physics) (New York: AIP) ISBN-13: 9781563960611 ISBN-10: 1563960613

Lidsey J E 2002 The Bigger Bang (Cambridge: Cambridge University Press) ISBN-13: 9780521012737 ISBN-10: 0521012732

Colkes P 2001 Cosmology: A Very Short Introduction (Oxford: Oxford University Press) ISBN-13: 9780192854162 ISBN-10: 019285416X

Smolin L 1999 The Life of the Cosmos (Oxford: Oxford University Press) ISBN-13: 9780195126648 ISBN-10: 0195126645

Hooper D 2007 Dark Cosmos: In Search of Our Universe's Missing Mass and Energy (Harper Perennial) ISBN-13: 9780061130335 ISBN-10: 0061130338

Chaisson E J 2002 Cosmic Evolution: The Rise of Complexity in Nature (Cambridge, MA: Harvard University Press) ISBN-13: 9780674009875 ISBN-10: 0674009878

Clark D H and Clark M D H 2004 Measuring the Cosmos: How Scientists Discovered the Dimensions of the Universe (Rutgers University Press) ISBN-13: 9780813534046 ISBN-10: 0813534046

Levin J 2003 How the Universe Got Its Spots: Diary of a Finite Time in a Finite Space (Anchor) ISBN-13: 9781400032723 ISBN-10: 1400032725

Croswell K 2001 The Universe at Midnight: Observations Illuminating the Cosmos (New York: Free Press) ISBN-13: 9780684859316 ISBN-10: 0684859319

Farrell J 2005 The Day Without Yesterday: Lemaitre, Einstein, and the Birth of Modern Cosmology (Basic Books) ISBN-13: 9781560259022 ISBN-10: 1560259027

Silk J 1997 A Short History of the Universe (W H Freeman) ISBN-13: 9780716760207 ISBN-10: 0716760207

Susskind L 2006 The Cosmic Landscape: String Theory and the Illusion of Intelligent Design (Back Bay Books) ISBN-13: 9780316013338 ISBN-10: 0316013331

Bartusiak M 2010 The Day We Found the Universe (Vintage Books) ISBN-13: 9780307276605 ISBN-10: 0307276600

Muller R A 2009 Physics for Future Presidents: The Science Behind the Headlines (W W Norton & Company) ISBN-13: 9780393337112 ISBN-10: 0393337111

Hawking S and Mlodinow L 2008 A Briefer History of Time (Bantam) ISBN-13: 9780553385465 ISBN-10: 0553385461

Glendening N K 2007 Our Place in the Universe (Singapore: World Scientific) ISBN-13: 9789812700698 ISBN-10: 9812700692

Goldsmith D and Goldsmith D 2000 The Runaway Universe: The Race to Find the Future of the Cosmos (Basic Books) ISBN-13: 9780738204291 ISBN-10: 0738204293

Close F 2009 Nothing: A Very Short Introduction (Oxford: Oxford University Press) ISBN-13: 9780199225866 ISBN-10: 0199225869

Kosso P 1997 Appearance and Reality: An Introduction to the Philosophy of Physics (Oxford: Oxford University Press) ISBN-13: 9780195115154 ISBN-10: 0195115155

Wilczek F 2010 The Lightness of Being: Mass, Ether, and the Unification of Forces (Basic Books) ISBN-13: 9780465018956 ISBN-10: 0465018955

Krauss L M 2013 A Universe from Nothing: Why There Is Something Rather than Nothing (Atria Books) ISBN-13: 9781451624465 ISBN-10: 1451624468

(Sample I) 2013 Massive: The Missing Particle That Sparked the Greatest Hunt in Science (Basic Books) ISBN-13: 9780465058730 ISBN-10: 0465058736

Close F 2013 The Infinity Puzzle: Quantum Field Theory and the Hunt for an Orderly Universe (Basic Books) ISBN-13: 9780465063826 ISBN-10: 0465063829

Mee N 2012 Higgs Force: The Symmetry-Breaking Force that Makes the World an Interesting Place (Lutterworth Press) ISBN-13: 9780718892753 ISBN-10: 0718892755

Chapter 17

Links to astronomy websites

This is a brief list of useful sites with Astronomical, Astrophysical, or Cosmological information. Some of these are just 'pretty pictures', but many are actually full of information at various technical levels.

Astronomy Resources: This is the best starting point if you want to branch out on your own.

Hubble Space Telescope Images: Wonderful images of current research in Astronomy and Astrophysics. Each comes with a caption and sometimes additional technical information.

Photographs by David Malin: Excellent photographs taken from the Anglo-Australian Observatory by probably the world's foremost astronomical photographer.

The Messier Gallery: A collection of images of objects in the Messier catalog.

The Supernova Cosmology Project at Berkeley and Harvard

COBE: The Cosmic Background Explorer: The initial observations of the anisotropy in the cosmic microwave background.

WMAP: The Wilkinson Microwave Anisotropy Probe: Modern measurements of anisotropy in the Universe as measured by the cosmic blackbody spectrum.

PLANCK: The European Space Agency microwave anisotropy program.

LIGO: The Laser Interferometer Gravitational-Wave Observatory (LIGO) project home page. This is a major effort to detect gravitational waves.

Sounds of Pulsars: This page has some recordings of pulsar "sounds". A suitably equipped workstation will allow you to listen to them.

doi:10.1088/978-1-6817-4100-0ch17

NRAO Home Page: Includes links to Green Bank, Kitt Peak, and the VLA radio telescope sites.

Areceibo Observatory Home Page

Jodrell Bank Center for Astrophysics

Caltech Astronomy Group: They operate Palomar Observatory, Keck Observatory, Owens Valley Radio Observatory, Caltech Submillimeter Observatory, Llano de Chajnantor Observatory, and the Big Bear Solar Observatory.

Berkeley Cosmology Group: Cosmology at Berkeley. CMB anisotropy, Supernova Cosmology Project, and much more.

Black Holes and Neutron Stars: Simulated trips to compact objects. Experience the effects of high gravitational fields. Good technical descriptions.

Stellar Evolution of a High-Mass Star: An animation showing the evolution of a high-mass star through the supernova stage to a radio pulsar.

Large Scale Structure: Images of large-scale structure from a number of re-shift surveys.

Images of Radio Galaxies and Quasars

Current Solar Images

Electronic Astronomy Textbook: An interesting project at the University of Oregon to create electronic teaching tools for Astronomy. The links are useful, including some to data for students to analyze.

Astronomy Notes: An Astronomy textbook on the Web.

Astronomy Picture of the Day

SkyView: A virtual telescope on the Web.

The Gemini Telescopes

The James Webb Telescope: The successor to the Hubble Space Telescope.

The Elegant Universe: Brian Greene's PBS series on String Theory.

The Fabric of the Cosmos: Brian Greene's second PBS series on Cosmology.

CPSIA information can be obtained
at www.ICGtesting.com
Printed in the USA
BVHW09s0732020918
526124BV00002B/5/P